Heuristic Search

Saïd Salhi

Heuristic Search

The Emerging Science of Problem Solving

Saïd Salhi
Centre for Logistics & Heuristic Optimisation
Kent Business School, University of Kent
Canterbury, United Kingdom

ISBN 978-3-319-84143-4   ISBN 978-3-319-49355-8   (eBook)
DOI 10.1007/978-3-319-49355-8

Printed on acid-free paper

This Palgrave Macmillan imprint is published by Springer Nature
The registered company is Springer International Publishing AG
The registered company address is: Gewerbestrasse 11, 6330 Cham, Switzerland

Preface

This monograph aims to provide an overview of heuristic search in general, to present the basic steps of the most popular heuristics and to stress the hidden difficulties as well as the challenging opportunities presented to researchers and practitioners when faced with complex combinatorial and global optimisation problems. Heuristic search, which is a combination of several attributes ranging from mathematical logic to experience, is an art and a science at the same time. Some of these heuristics are inspired from natural and biological evolution, artificial intelligence and even music. Their applications are now widely used in engineering, finance, management, sport and medicine among others.

This book is written in an informal way so that students and non-specialists will find it easy to read and will get interested in this exciting decision science-based subject. Throughout the text, comments and observations will be added to keep an open mind on how these approaches are conceived, analysed and applied. Each chapter starts with a brief introduction and finishes with a small summary to provide the reader with the opportunity to read the chapter in its entirety, or just the overview.

At first glance, these methods present themselves as simple and easy, but once the experience kicks in, the researcher and/or practitioner will start to appreciate that their implementation is harder than expected, challenging and probably addictive, if you permit me to use this word. By the end of this book the reader will acquire the basic philosophy of these exciting and lively methodologies, and a strong foundation to build on when designing and analysing his/her own techniques.

It is also worth mentioning that heuristic search is considered as a grey research area that is not as structured as classical research, and hence considered by some as less rigorous. Meanwhile, others like myself take the opposite view of embracing it due to its flexibility and soft boundaries that make the science even more exciting and rewarding.

This book is suitable as an introductory and general book on heuristic search for combinatorial and global optimisation for research students at the master's and doctoral levels in Operational Research, Computer Science, Engineering and Medicine. Consultants and analysts in problem-solving and decision-making will also find this book to be a practical addition to their analysis tool kit while being useful and easy to read and understand.

Centre for Logistics & Heuristic Optimisation   Saïd Salhi
Kent Business School
University of Kent
Canterbury, UK

Acknowledgements

I would like to thank all my 25 former PhD students for their invaluable contribution which improved directly or indirectly the content as well as the presentation of the present manuscript. I would also like to thank my colleagues working in the area of heuristics irrespective of which disciplines they belong to. Their comments and advice were much appreciated, in particular, Professors Jack Brimberg, Nenad Mladenović and Jonathan Thompson. I am also grateful to my current PhD student Jeeu Fong Sze (Yvonne) for her technical support.

I would like to dedicate this book to my parents who have respected my wishes through hard times and to my beloved wife Pauline for her patience and hard work in keeping our family of five lovely children (Riadh, Tarek, Louisa, Nassim and Salim) safely under one roof.

Contents

List of Algorithms

List of Figures

List of Table

1

Introduction

1.1 Introduction

Many real-life applications can be modelled reasonably well and solved optimally by one of the classical optimisation techniques familiar to the Operational Research/Management Science/Computer Science/Engineering communities such as linear programming, integer programming, non-linear programming, dynamic programming (DP) and network-based methods, among others. However, there are a lot of applications where the combinatorial effect of the problem makes the determination of the optimal solution intractable, and hence these standard optimal (or exact) techniques become unsuitable. This is because the computer time needed to find such a solution could be too large to be acceptable in real life or the solution found is not really the overall best (global optimum) but just one of the local optima that can be relatively poor when compared to the overall (or global) best one. To overcome such a drawback, heuristic methods were devised with the aim to provide the user with reasonably good solutions, if not the best.

Although heuristic methods <u>do not guarantee optimality</u>, in some situations, they seem to be the <u>only way forward</u> to produce concrete

© The Author(s) 2017
S. Salhi, *Heuristic Search*, DOI 10.1007/978-3-319-49355-8_1

results. To date, heuristic search methods have been widely used in business, economic, sports, environment, statistics, medicine and engineering problems that were found difficult to solve otherwise. The view of adopting heuristics to solve approximately a large spectrum of complex optimisation problems that were not possible to solve before is now much more accepted and welcomed. There are however measures of performance to assess these heuristics which I shall briefly discuss later in this chapter.

The word 'heuristic' originates from a Greek word that means discover and explore in the wider sense. Heuristics are also known as approximate techniques. The main goal in heuristic search is to construct a model that can be easily understood and that provides *good* solutions in a *reasonable* amount of computing time. Such techniques consist of a combination of scientific components such as mathematical logic, statistics and computing, as well as human factors such as experience, and also in many cases a good insight of the problem that needs to be addressed. The latter may give the impression that such research development can be relatively restrictive when compared to the general techniques, but in my view, this is a crucial component in the design of a heuristic as this could make the technique much faster and more relevant to the problem under study. In certain cases, this can even lead to new ideas that would not have been thought about otherwise. These new ideas can sometimes be translated into formal rules and algorithms that can be used for a wider class of related applications and, hence, could be very rewarding.

In the remainder of this chapter, I first give a brief methodology on how to approach a real problem followed by some performance measures for the evaluation of a given method, and then I describe the proposed simple heuristic categorisation. The main focus of global search is to approximately try to avoid being trapped into a poor local optimum. This is performed by either (a) accepting only improving solutions (Chap. 2), (b) accepting certain uphill moves (Chap. 3) or (c) providing several solutions at once which when combined somehow may generate some new better solutions (Chap. 4). The recent area of hybridisation between heuristics or between heuristics and exact methods is discussed in Chap. 5 followed by some implementation issues in Chap. 6. The final chapter covers real-life applications as well as those applications commonly used by academics, followed by the last section that summarises the conclusions and some of the research avenues that I believe to be challenging.

1.2 The Optimisation Problem

A general optimisation problem for the case of minimisation can be defined in the following form:

$$(P) : \begin{cases} \text{Minimise} & F(X) \\ st & X \in S, S \subseteq E \end{cases}$$

In many applications, (P) can be rather complex to solve because of the following:

(i) The solution space E can be either a finite and very large set rendering (P) as combinatorial optimisation problem, or $E = \mathbb{N}^n$ making (P) an integer optimisation problem, or $E = \mathbb{R}^n$ a continuous optimisation problem.

(ii) The decision variables, X, can be continuous, integer, binary or any combination of the above types.

(iii) The objective function, $F(.)$, may not be necessarily linear, continuous or even convex. $F(.)$ can also be made up of several objectives, some of which may be conflicting.

(iv) The feasibility set, S, may not be necessarily convex and may even include disconnected subsets.

(v) And finally, the viability of the parameter values within the definition of F and S can be critical as these may be probabilistic, estimated or even not fully known.

In brief, when set S is a discrete set, the problem (P) falls into the category of discrete optimisation problems (also known as combinatorial optimisation), whereas if it is a continuous set, (P) is referred to as a continuous optimisation problem (also known as global optimisation).

Local versus Global Optimality

Consider $X \in S$ and let $N(X) \subset S$ be a given neighbourhood of X. $N(X)$ could be defined by a 'small' area around X. Some definitions of the neighbourhoods will be defined later depending on the application.

Local minimum (maximum)
  $\widetilde{X}$ is a local minimum (maximum) with respect to neighbourhood $N(.)$
  if $F\left(\widetilde{X}\right) \leq (\geq) F(X) \forall X \in N(X)$
Global minimum (maximum)
  X^* is a global minimum (maximum) if $F\left(X^*\right) \leq (\geq) F(X) \forall X \in S$

For instance, if Ω represents all the possible neighbourhoods and Λ the set of all local minima (maxima), X^* can also be defined as $X^* = \text{Arg M}$ in $\{F(X); X \in \Omega(X)\}$ or $X^* = \text{Arg Min}\left\{F\left(\widetilde{X}\right); \widetilde{X} \in \Lambda\right\}$.

In brief, the global minimum (maximum) X^* is the local minimum (maximum) that yields the best solution value of the objective function F with respect to all neighbourhoods.

Local Search

This is the mechanism (i.e., operator or transformation) by which $\widetilde{X}$ is obtained from X in a given neighbourhood $N(X)$. In other words, $\widetilde{X} = \text{ArgMin}\{F(X); X \in N(X)\}$.

1.3   Possible Methodological Approaches

Despite the above difficulties, many applications fall into suitable formulations where optimality can be guaranteed (e.g., optimal solutions can be determined). In this book, we concentrate on those cases where an optimal solution cannot be guaranteed while a 'good' solution, which may or may not be optimal, is required for practical purposes.

Most general optimisation methods fall into two main categories known as exact (optimal) algorithms and approximate or heuristic algorithms. The former is capable, at least theoretically, to guarantee the optimal solution in a finite number of steps, whereas the latter category includes heuristics. These are based on a set of rules which can be developed from experience, problem characteristics, common sense and mathematical logic. They have the tendency to tackle the problem on its entirety and in a reasonable amount of computer time. The solutions produced by this class of methods are, unfortunately, not necessarily optimal. However, the performance of these methods can be evaluated using certain performance criteria which will be briefly highlighted in subsequent sections of this chapter.

It is also worth noting that it may be useful to distinguish between heuristic and approximation algorithms as both aim to yield a feasible solution. The only difference is that the approximation methods have a certain guarantee of the solution quality based on the worst case scenario. For instance, for the travelling salesman problem (TSP), the well-known Christofides' algorithm which is based on solving the minimum spanning tree has a worst quality ratio of 1.5 (cost of the approximate TSP tour/cost of the optimal TSP $\leq$ 1.5). Also, a simpler approximate method in bin packing is the next fit (NF) algorithm where the ratio is 2. This information, though theoretically interesting, needs to be carefully translated to the user as it could unfortunately be wrongly understood by the user. Most heuristics on the other hand, have no similar mathematical bounds for quality purposes though some special ones may have, and some researchers are currently attempting to find them. There are other useful ways for evaluating the quality of a given non-optimal approach including heuristics which I shall revisit later in this chapter.

When approaching complex real-life problems, I consider the following four commonly applied steps to be followed in the sequence given below. Note that this ordered list is not exhaustive as other combinations do exist though these may be relatively more difficult to define explicitly.

The aim is to apply

1. an exact method to the exact (true) problem; if not possible go to (2).
2. a heuristic method to the exact problem; if not possible go to (3).
3. an exact method to a modified (approximated) problem; if not possible go to (4).
4. a heuristic (or approximate) method to the approximated problem.

Though these rules are presented in the above ranking, the complexity in the design of the heuristic in Step (2) which aims to retain the true characteristics of the problem, and the degree of modification of the problem in Step (3) are both crucial points when dealing with practical problems. It may be argued that Steps (2) and (3) could swap places. The idea is to keep the characteristics of the problem as close as possible to the true problem (small simplification only) and then try to implement (1) or (2). The following recursive process, as given in Algorithm 1.1, can be adopted when approaching complex real-life problems.

Algorithm 1.1: A Possible Recursive Methodological Approach

Step 1   Define (*P*) as the current problem.
Step 2   Choose a suitable exact method to solve (*P*).
          If it is possible, then conduct the experiment and stop; otherwise go to Step 3.
Step 3   Choose a suitable heuristic to solve (*P*).
          If it is possible, then conduct the experiment and stop, else modify (*P*) and go back to Step 1.

The level of modification ought to be carefully considered. A massive modification may make the problem very simple to solve, but will have little resemblance to the original problem, whereas a tiny modification may lead to a problem that could still be hard to solve.

Another related approach would be to start with a simplified or a relaxed version of (*P*), find a solution and check whether those complex constraints are satisfied. If it is the case, there is no need to worry about the original problem as the solution is optimal. If some constraints are violated, which is likely to happen especially at the beginning of the search, additional characteristics are then introduced gradually. The process is repeated until the problem becomes impractical to solve, and hence the feasible solution found at the previous stage can then be used as the final solution of the problem. Note that this concept is the basis for dual-based methods such as the dual simplex method in linear programming, and also in Lagrangean Relaxation, a topic I shall visit in the hybridisation chapter.

In summary, we can say that the modifications can be implemented at three levels, namely, the input stage (problem characteristics), the algorithm stage (whether it is an exact or a heuristic method) and, finally, the output stage (solutions found). The choice of whether to concentrate on the input, the algorithm or the output is critical. I think one way forward is to try to retain the input to its originality as much as possible and to focus on the other two. Obviously if slight changes made to the input lead to a problem that can be solved optimally or efficiently, such an approach should not to be discarded.

The hierarchy presented in Algorithm 1.1 is used to emphasise the need to maintain the problem characteristics as close as possible to the ones of the *true* problem. This concept is highlighted by the following observation that *it is better to have a good and acceptable solution to a true problem rather than an optimal solution to a problem that has very little resemblance to the original problem (e.g., solving the wrong problem).*

The lack of understanding, and in some cases, the lack of appreciation of the difficulties of the real-life problem, could lead to a less favourable relationship between the two main parties, namely, the end user (company) and the consultant or researcher. This unfortunate scenario may encourage practitioners to distant themselves from academics, which, in turn, may lead to the least favourable outcome for either party. This trend is becoming less and less strong as nowadays the relationship between universities and the outside world steadily improves as mutual benefits are now being generated and measured. The company gains an added competitive advantage, while the academic enhances his/her research portfolio. Furthermore, for example, in the UK, the research impact in practice is now used towards the overall research assessment score for the school or faculty involved as required by the Research Excellence Framework; see REF (2014).

1.4   Need for Heuristics

I would like to stress once more that heuristics are used only when exact methods, which guarantee optimal solutions, are impractical because

(a) either the computational effort required is excessive, or
(b) the risk of being trapped at a local optimum is too high.

Given the above aspects, in such circumstances, heuristics become virtually the only way to help the user find reasonably acceptable solutions. The reasons for accepting and promoting heuristics include the following with some being noted in Salhi (1998, 2006).

1. Heuristics can be the only way forward to produce concrete solutions to large and difficult combinatorial and global optimisation problems.
2. They are easily adaptable and accessible for additional tasks or constraints if necessary.
3. Heuristics can be easily supported by graphical interface which can help the user in understanding better the progress of the search, and in modifying the outcomes if need be, hence avoiding the drawbacks of black box products.
4. Management and less specialised users find them reasonably easy to understand and therefore to interact with.
5. These methods are not difficult to code, validate and implement.
6. Management can introduce some unquantifiable measures indirectly to see their effect as solutions can be generated while conducting an interesting scenario analysis.
7. These methods are suitable for producing several solutions and not only a single one, thus providing the user with the confidence, flexibility and the opportunity to choose one or more solutions for further investigation if need be.

8. The design of heuristics can be considered as an *Art* since a proper insight of a problem is fully required, but also as a *Science* as it requires some form of logic. This interesting combination can bring the user and the researcher closer together at the design stage, enabling both parties to contribute positively to the success of the project.

1.5 Some Characteristics of Heuristics

The last item fits well within the philosophy of science as presented by Popper (1959) who emphasised that the design of heuristics ought to favour insight over efficiency, and also that learning is part of a process of exploiting the failures of heuristics.

The following characteristics are worth considering in the design of a given heuristic though these do not have to be followed exactly. Some are by-products of the attributes that make heuristic necessary as mentioned in the previous section, whereas some are added for generalisation purposes. For simplicity, these characteristics are only summarised below while noting that more details can be found in Laporte et al. (2000), Salhi (1998, 2006) Hansen et al. (2010) among others:

(i) Simple and coherent—the method should follow well-defined steps.
(ii) Effective and robust—it should be reliable enough to provide good or near optimal solutions irrespective of the instances used.
(iii) Efficient—the time required needs to be acceptable taking into account whether the problem is strategic or tactical. Either way the heuristic ought to be implemented in an efficient way as will be highlighted in the implementation chapter.
(iv) Flexible—this needs to cater for modification and adaption. For instance, the addition/deletion or modification of some steps can be easily accommodated to include new ideas and to be able to solve related problems while retaining the other strengths of the heuristic. This obviously includes the flexibility in allowing interaction with the user as he/she is the one who takes control and has the final decision.

Heuristics, in general, including global search heuristics (also called modern heuristics, metaheuristics or meta strategies) that shall be presented throughout this book have the properties to exploit the overall region rather than individual or local regions. However, two natural questions arise:

(a) Is there any guarantee that escaping from a local optimum will turn into exploring new and more promising regions of the feasible set?
(b) Do such regions exist or do we expect the local minima (maxima) to be uncorrelated?

The answers of these questions are far from being definite but provide a platform to base the search when designing a heuristic. Theoretically the answer to (a) is no, but the principle of getting closer to a yes could be helpful when constructing diversification strategies. For (b) the local minima are theoretically uncorrelated in general though some local optima may not be far from each other.

Metaheuristics

Note that metaheuristics (also known as modern heuristics) are considered as higher level heuristics that are devised to guide other constructive heuristics or local searches to reduce the risk of being trapped into a poor local optimum. Though this definition helps in differentiating between low-level heuristics, which could include, for example, constructive and simple heuristics as well as local searches, and higher level ones that aim to control the lower ones, in this monograph I shall refer to the word heuristics whenever possible irrespective of their level. The aim is to keep the terminology simple while providing freedom and flexibility to the word heuristics as it was historically described. In other words, it relates to mechanisms used to discover and search, using ways that may be simple (low-level heuristics or local search) or more complex and powerful (metaheuristics and others to mention later in this book).

1.6 Complexity and Performance of Heuristics

Heuristics need not be linked to the phrase usually known as 'quick and dirty' or 'guess work'. Heuristics are wider in concept than that those simple phrases though in some situations simple implementations may do the job.

Heuristics ought to be carefully devised to represent the full characteristics of the problem (not necessarily the generalised problem), and validated and tested from both points of view, namely, computing time and solution quality. The main criteria for evaluating the performance of a new heuristic may be classified under two headings, namely, the quality of the solutions provided and the computational effort, measured in terms of computer processing units (CPU) time. Other criteria such as simplicity, flexibility, ease of control, interaction and friendliness can also be of interest, and more particularly to the user (see Section 1.5). For further details, see Reeves (1995), Salhi (1998, 2006), Barr et al. (1996), Johnson (1996) and Laporte et al. (2000).

Solution Quality

There are at least five performance measures required to test a given heuristic. These include empirical testing, worst case behaviour, probabilistic analysis, lower bound (LB) solutions and obviously benchmarking.

(a) *Empirical testing*- this can be based on the best solutions of some of the existing heuristics when tested on a set of published data. Here, we can produce average deviation, worst deviation, the number of best solutions and so on. This measure, which is one of the most useful and commonly used approaches in practice, is simple to apply and can be effective when published results exist. Although accuracy is guaranteed as the results obtained are known with certainty, this type of analysis can provide only statistical evidence about the performance of the heuristic for other not yet tested instances.

(b) *Worst case analysis*—a pathological example, which is represented purposely to show the weakness of the algorithm, needs to be constructed. It is usually very hard to find such an example especially if the problem is complex. One of the drawbacks of such an analysis, though theoretically strong, is that in practice the problem under study rarely resembles the worst case example. It is however reassuring that at least such a heuristic produces in the worst scenario solutions that are $\alpha\%$ inferior at most. However, this information can also be misleading, and could put the user off if such an information is not communicated appropriately. One way is to understand the problem rather well and then construct a worst case example for a class of problems sharing similar characteristics to the real problem. Such a result provides a useful measure and a guaranteed performance which is not too far away from the real story. This measure is similar to the quality bounds usually developed for approximation algorithms (first fit decreasing [FFD] for bin packing, Christofides spanning tree based for the TSP, etc.).

(c) *Probabilistic analysis*—the density function of the problem data needs to be determined, allowing statistical measures to be derived such as average and worst behaviour.

(d) *Lower bounds*—one way is to solve a relaxed problem where many of the difficult constraints are removed (e.g., LP relaxation), or where the transformed problem falls into a nice class of easy problems (Lagrangean relaxation). The main difficulty is that the LB solutions obtained have to be rather tight to tell the quality of the heuristic solution; otherwise, misleading conclusions could be drawn.

(e) *Benchmarking*—in situations where a benchmark solution already exists, which is given by the user, the obvious way is to see how the solution found by the new heuristic compares with the benchmark. This can have positive impact in both the design of the heuristic (as initial results may not be necessarily competitive and hence enhancements could be added making the heuristic more efficient and powerful) and the way the user conceives the result. This is a win-win situation because if the results are better, the company could have a competitive advantage over its competitors while learning where improvements can be made. But, if the results happen to be the same, this could also demonstrate that what the company is doing is good as it is now formally tested by an outsider. This may lead to a growth in confidence and self-belief for the user besides gaining additional trust from senior management.

The relationship between the user and the analyst is of crucial importance at this level especially if the user is aware of the primary runs of the heuristic as discussed in the earlier point (e). Inferior or infeasible solutions can send the wrong signal to the user, and therefore, a good understanding on the progress of development of the heuristic is vital. This extra human effort does not only avoid communication hick-ups from escalating, but also helps in building a friendly atmosphere in which modification and improvement are part of the design that needs to take place. In some cases, the initial runs are not shown to the user, and only when positive results are found will the user commence to inject feedback. The former strategy can be more effective if it is well looked after, whereas the latter one is good as long as the user is not supposed to react negatively in early stages of the work. A strategy that sits in between the above two ideas may be the most appropriate in many cases; however, the breaking point is dependent on the analyst's relationship with the user, the company's reputation, the routine checks which are initially arranged between the user and the analyst and the urgency by which the end product needs to be delivered.

Computational Effort

Computational effort is usually measured by the time complexity and the space complexity of a solution procedure. The former describes the computing time the method requires for a given instance, whereas the latter measures the storage capacity needed when solving a given instance. Unfortunately, the latter is seldom discussed in the literature.

Time Complexity

The time complexity of an algorithm is defined by $O(g(n))$ where n denotes the size of the problem.

If $g(n)$ is a polynomial function of n (e.g., $an^k, k \geq 1, a > 0$), then the problem can be solved within a reasonable amount of computation time. But if $g(n)$ is an exponential function of n, the problem may be difficult to solve. Such types of problems are usually known to be NP hard, see Garey and Johnson (1979) for more details.

For instance, in bin backing, the aim is to minimise the number of bins of equal volume or weight to be used where the weight of each of the n items to be stored in the bins is known. The FFD heuristic has a runtime $O(n^2)$, whereas the NF heuristic is relatively quicker having a runtime $O(n)$. However, FFD has a tighter worst quality ratio of $\frac{N_{FFD}}{N^*} < \frac{11}{9} + \frac{4}{N^*}$, whereas $\frac{N_{NF}}{N^*} \leq 2$, with N^* representing the minimal number of bins and N_H the number of bins found by heuristic H.

Space Complexity

Although this issue is less referenced when compared to time complexity, the way the data are stored and retrieved is an important issue in heuristic design. An efficient data handling not only uses the smallest necessary storage capacity in the computer but it can also save a large amount of computing time by not calculating unnecessary information which is either redundant or already found in earlier iterations. The issue of memory excess is also well known when using commercial optimisation software such as CPLEX, ILOG, LINDO, Xpress-MP, GuRobi and so on. The problem may not be resolved because of the computing time required, but simply because the computer runs out of memory. A massive storage space may be needed for the software to sometimes even start the optimisation or during the search due to the large amount of branching within Branch and Bound (B&B). However, there can sometimes be ways around the problem. For instance, in some cases, using another branching strategy within CPLEX such as depth first instead of best first helps in reducing the storage burden up to a certain point.

'Real' Meaning of Large or Small Computing Time

It is worth highlighting that the concept of large or small computer time ought to be relative to both the nature of the problem (strategic, tactical or operational) and the availability of the computing resources. The time for interfaces is usually ignored in research though it can constitute an important part of the total computing time in practice. This additional time could however be taken to be a constant if carried out by professionals in software engineering or related areas.

The impact of computing effort is directly related to the importance of the problem. For instance, if the problem needs to be solved once or twice a day (very short-term planning), it is essential that the algorithm be quick, whereas if the problem is solved at a medium or at a strategic level (once every month or year), it becomes less important to give a high priority to the CPU time, but more so to the quality of the solution. Consider the case where the problem requires a large investment such as with the location of new facilities, the purchase of expensive equipment, planning of the workforce, among others. In these instances, it does not matter so much if the method takes ten minutes, five hours or even a day, as long as good solution is provided at the end. However, one may also argue that the algorithm should not be too slow for simulation purposes even when tackling strategic problems, so several options and scenarios could be investigated providing the decision makers with additional flexibility. For instance, it may be that the third best solution, when taking other factors into account (i.e., unquantifiable or even confidential aspects) besides the cost, becomes the most practical option to choose from all available solutions.

A good and efficient computer code can obviously save a lot of unnecessary computing time. This can be achieved by avoiding redundant calculations of already computed full or partial information. A good data structure (DS) which keeps track of already computed information could also help. The reduction in computing time can also be obtained by introducing some reduction tests (neighbourhood reduction) which eliminate testing certain cases (or combinations) which in the analyst's view are unlikely to influence the final best solution. There will generally be a trade-off between speed by which the best solution is obtained and the quality of that solution. I shall visit this important aspect when discussing the implementation issues in Chap. 6.

1.7   A Possible Heuristic Classification

There are several ways to classify heuristics including the following:

(i)   Deterministic vs stochastic
(ii) One solution at a time vs several solutions taken simultaneously (population)
(iii) Fast and dirty vs slow and powerful
(iv) Classical heuristics vs modern heuristics

In this monograph, the following categorisation consisting of four groups that are not necessarily entirely disjoint, as may be conceived by some researchers, is provided:

(i)   Improving solutions only
(ii) Not necessarily improving solutions
(iii) Population-based
(iv) Hybridisation

The first two groups are completely disjoint, whereas the last two could interrelate with each other as well as with the first two. For clarity of presentation, such links are highlighted in Fig. 1.1 in dash, whereas the main links are shown in bold.

1.8   Summary

A general view of heuristics is given followed by the need for their usage in practice. As these methods are not exact, some performance measures and appropriate characteristics are provided to guide the user when designing such techniques. A simple classification of heuristics is also provided that

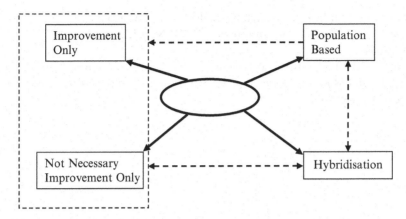

Fig. 1.1 A heuristic classification

will be used throughout this monograph. In the next chapter, I shall concentrate on the first category, namely, improvement-only heuristics.

References

Barr, W. S., Golden, B. L., Kelly, J. P., Resende, M. G. C., & Stewart, W. R. (1996). Designing and reporting on computational experiments with heuristic methods. *Journal of Heuristics, 1*, 9–32.

Garey, M. R., & Johnson, D. S. (1979). *Computers and intractability: A guide to the theory of NP-completeness.* San Francisco: Freeman.

Hansen, P., Mladenović, N., Brimberg, J., & Moreno Perez, J. A. (2010). Variable neighbourhood search. In M. Gendreau & J. Y. Potvin (Eds.), *Handbook of metaheuristics* (pp. 61–86). London: Springer.

Johnson, D. S. (1996). A theoretical guide to the experimental analysis. *AT&T Research Report, Bell Laboratories, University of Michigan.*

Laporte, G., Gendreau, M., Potvin, J.-Y., & Semet, F. (2000). Classical and modern heuristics for the vehicle routing problem. *International Transactions in Operational Research, 7*, 285–300.

Popper, K. (1959). *The logic of scientific discovery.* London: Hutchinson.

Reeves, C. R. (1995). *Modern heuristic techniques for combinatorial problems.* Oxford: Blackwell Scientific Publications.

REF. (2014). REF 2014: Research Excellence Framework. http://www.ref.ac.uk

Salhi, S. (1998). Heuristic search methods. In G. A. Marcoulides (Ed.), *Modern methods for business research* (pp. 147–175). New Jersey: Lawrence Erlbaum Associates.

Salhi, S. (2006). Heuristic search in action: The science of tomorrow. In S. Salhi (Ed.), *OR48 Keynote papers* (pp. 39–58). Birmingham, UK: ORS Bath, The OR Society.

2

Improvement-Only Heuristics

2.1 Neighbourhood Definition and Examples

I first present the basic definition of neighbourhood followed by some simple combinatorial type examples that are commonly used in the Operational Research/Computer Science literature.

Neighbourhood Definition A solution, say $X \in S$, has an associated set of neighbours, say $N(X) \subset S$, which is called the neighbourhood of X. A solution, $X' \in N(X)$, can be obtained directly from X by an operation which is referred to as a move (or transition m), say $X' = m(X)$. Such a move can be a single move consisting of simple or complex operations, or a series of moves put together that can be used either in a well-defined and deterministic sequential manner, or in a random or a pseudo-random way.

The Travelling Salesman Problem

For instance, for the case of the Travelling Salesman Problem (TSP) where the aim is to get the least cost tour that originates say from a

© The Author(s) 2017
S. Salhi, *Heuristic Search*, DOI 10.1007/978-3-319-49355-8_2

home town (a depot or a facility), visits each city only once and returns back to the home town. The idea is to have a sequence of nodes (cities, say $i = 0, 1, \ldots, n$) starting from the home town say 0. Let d_{ij} denote the distance between city i and city j where $i, j = 0, 1, \ldots, n$.. Let $\sigma = (0, \sigma_1, \ldots \sigma_n, 0)$ be the sequence where σ_j denotes the city occupying the jth position in the tour and $F(\sigma)$ the total travel distance (cost) of the tour $\sigma \in \Omega$ where Ω represents all the possible sequences. The objective is to find the sequence with the least total distance (i.e., find $\sigma^* = \text{ArgMin}\{F(\sigma), \sigma \in \Omega\}$).

One possible neighbourhood for the TSP would be to exchange the position of a city with one of its neighbours. For example, the position of the ith and the jth cities in σ could be swapped leading to σ' as follows:

$$\sigma = (0, \sigma_1, \ldots, \sigma_i, \ldots, \sigma_j, \ldots, \sigma_n, 0) \rightarrow \sigma' = (0, \sigma_1, \ldots, \sigma_j, \ldots, \sigma_i, \ldots, \sigma_n, 0).$$

The p-Median Problem

Suppose we have n customers that need to be served from p facilities. The aim is to find the location of these p facilities out of m potential sites such that the average distance (or the total distance) from these p open facilities to all customers is minimised. Let the facility configuration of these p facilities be $\sigma = (\sigma_1, \ldots, \sigma_p)$ where $\sigma_j \in \{1, \ldots, m\}$ represents the jth chosen facility and Ω being the set of possible configurations. The objective is to choose the configuration $\sigma^* = \text{ArgMin}\{F(\sigma), \sigma \in \Omega\}$ where

$$F(\sigma) = \sum_{i=1}^{p} \sum_{j \in A_i} d_{j\sigma_i} \text{ with } A_i = \left\{ j \in \{1, \ldots, n\} \mid d_{j\sigma_i} = \text{Min}\left(d_{j\sigma_k}; \sigma_k \in \sigma\right) \right\}$$

For the p-median problem, a possible neighbourhood would be to close an open facility say $\sigma_i \in \sigma$ and choose a potential site from the non-selected ones, say $k \notin \sigma$. In brief, we moved from σ to its neighbour σ' as follows

$$\sigma = (\sigma_1, \ldots, \sigma_i, \ldots, \sigma_p) \rightarrow \sigma' = (\sigma_1, \ldots, k, \ldots, \sigma_p).$$

In other words, $N(\sigma) = N(\sigma) - \sigma_i \cup k$ where $\sigma_i \in \sigma$ and $k \notin \sigma$.

The Bin Packing Problem

Consider there are n items of weight $w_j; j = 1, \ldots, n$ and bins of equal weight (can be different as well) W. The aim is to minimise the number of bins such that all the items are included. Assume that we have already obtained p sequences (bins) found by one of those approximation algorithms such as the FFD. We consider $p > LB$ with the lower bound LB taking the smallest integer that is larger than $\dfrac{\sum_{j=1}^{n} w_j}{W}$, otherwise the solution is obviously optimal. Let $B_k = \left\{ B_k^1, \ldots, B_k^j, \ldots, B_k^{N_k} \right\}; k = 1, \ldots, p$ be the kth bin with B_k^j representing the jth item in bin k and N_k being the number of items in bin k. Let $W_k = \sum_{j=1}^{N_k} w_{B_k^j}$ be the total load in bin k. If infeasibility is not allowed, one possible move is to shift one item from one bin to another as long as the total weight of the receiving bin is not violated. For instance, remove item B_k^j from bin k and inserting it in bin l if $W_l + w_{B_k^j} \leq W$. This process is repeated until one bin is emptied.

The Vehicle Routing Problem

When there is a restriction on either the total travel time or/and the maximum total load to be transported on a vehicle (truck), the TSP may no longer be valid as the tour may use a larger load than the vehicle capacity say Q and could take longer than the daily travel time, say T. Consider there are n customers that need to be served from the depot which we refer to by 0. Each customer has its own demand say

$q_j; j = 1, \ldots, n$. Each vehicle has the same capacity Q and could not travel more than T (say nine hours per day including break). Here, all vehicle routes start and finish at the same depot. The aim is to find the number of vehicles used with their respective sequences (vehicle routes) so that the total distance travelled is minimised. Note that when Q and T are both very large, the problem reduces to the TSP with the depot as the home city.

Consider we have p routes and $R_k = \left(0, R_k^1, \ldots, R_k^i, \ldots, R_k^j, \ldots, R_k^{N_k}, 0\right)$; $k = 1, \ldots, p$, with R_k^j denoting the jth customer in the kth route and N_k being the number of customers in route k. Possible neighbourhoods based on R_k include

(i) swapping the places of the ith and the jth customers (say R_k^i with R_k^j) in R_k to obtain say $R_k' = \left(0, R_k^1, \ldots, R_k^j, \ldots, R_k^i, \ldots, R_k^{N_k}, 0\right)$,

(ii) moving customers between routes say customer R_k^i from R_k is moved to R_l and customer R_l^j from R_l to R_k making the two new routes R_k' and R_l' $R_k' = \left(0, R_k^1, \ldots, R_l^i, \ldots, R_k^j, \ldots, R_k^{N_k}, 0\right)$ and $R_l' = \left(0, R_l^1, \ldots, R_k^i, \ldots, R_l^j, \ldots, R_l^{N_l}, 0\right)$; $l \neq k$ as long as capacity and time constraints are not violated. The exchange does not need to be using the same customer position as shown here.

Non-linear Optimisation Problems

The classical non-linear optimisation problem (case of minimisation) can be formulated as

$$(P): \begin{cases} \text{Minimise} & F(X) \\ st & X \in S, S \subseteq R^n; X = (x_i), \ x_i \in [L_i, U_i], i = 1, .., n \end{cases}$$

The neighbourhood can be expressed as $N(X) = \left\{X' \in S \quad \text{such that} \quad d(X, X') \leq R\right\}$ where R is the radius of the neighbourhood defined either as threshold or by upper and lower bounds of the coordinates $x_j; j = 1, \ldots, n$. One way would be to take x_j from the incumbent solution $X = (x_i)_{i=1,\ldots,n}$ and replacing it by its neighbour $x'_j \in [L_j, U_j]$ to obtain X' as follows:

$$x'_j = \begin{cases} L_j & \text{if } x_j - \alpha(U_j - L_j) \leq L_j \\ U_j & \text{if } x_j + \alpha(U_j - L_j) \geq U_j \\ x_j - \alpha(U_j - L_j) & \text{otherwise} \end{cases}$$

In other words, $X = (x_1, \ldots, x_j, \ldots, x_n) \rightarrow X' = (x_1, \ldots, x'_j, \ldots, x_n)$ with $\alpha \in [0, 1]$ randomly generated.

If the problem is further constrained, feasibility can then be imposed while generating x'_i or restricting it marginally while attaching some penalties to the violation as part of an augmented objective function. This is used as part of the Lagrangean relaxation which will be discussed in Chap. 5 and also embedding in other heuristics so their search could explore a wider search space.

2.2 Basic Descent or Hill Climbing Method

The main steps of the basic descent method (BDM) are summarised in Algorithm 2.1. BDM has two main components namely the generation of the initial solution and the way such a solution is improved via the selected neighbourhood $N(.)$. An efficient implementation and a good design of these two components are worth the effort as these will contribute to the final quality of the solution.

Algorithm 2.1: The Basic Descent Method

Step 1   (*Initial Phase*) Select an initial solution, say $X \in S$.

Step 2   (*Improvement Phase*) Choose a solution $X' \in N(X)$ such that $F(X') < F(X)$.

If there is no such X', X is considered as a local minimum and the method stops, else set $X = X'$ and repeat Step 2.

Initial Solutions (Step 1)

An initial solution can be generated either randomly or constructed using simple rules such as those constructive type heuristics that build the solution piece by piece until the final solution is completed. At each step, the inclusion of the new piece is performed by certain selection rules some of which are relatively more computationally intensive than others. As an example consider the case of the TSP.

Construction-Based Heuristics for the TSP

The following rules are commonly used.

(i) Randomly keep adding the cities one by one until all cities are visited. This is a pure random rule which may produce very poor quality. One way would be to repeat the process several times and select the tour with the least total cost.

(ii) Start at the home city, visit the nearest city not visited yet and so on until all cities have been visited then come back to the home city. Let σ denote the ordered sequence of the nodes already selected on the tour. Initially $\sigma = \{0\}$ then

$$\sigma = \sigma \cup \{k\} \quad \text{such that } k = \operatorname{argMin}\{d_{ij}, j \in \sigma\} \tag{2.1}$$

where i represents the last point on the partial constructed tour.

(iii) Put the cities in their polar coordinates with respect to one given axis, and sort the cities based on their angles. The solution is made up by following the list until the final city is reached which is then connected to the first one. This is similar to (ii) except the distance is replaced by the angle.

(iv) Insertion-based rules—starting from the home city, choose the first city (the nearest city or furthest, random), then select the next city using the least insertion cost. For instance, node k will be selected and inserted between $i - 1$ and i. This process is repeated until all

cities have been visited. Initially, $\sigma = \{0, i_1\}$ where i_1 is the first city to be added to the home city, 0.
$\sigma = \sigma \cup \{k\}$ such that
$$k = \text{argMin}\{d_{i-1j} + d_{ij} - d_{i-1,i}; \quad j \notin \sigma; i - 1, i \in \sigma\}$$

Basic Enhancement of Step 1 in BDM

One possible approach is to build a powerful but efficient constructive heuristic that provides good quality solutions which can then be fed into the improvement step of the BDM. It is worth noting that the above rules given in Step 1 can be short sighted in their search. One way would be to incorporate long-term strategies based on a look-ahead scheme. Here, two or three intermediary mini steps are considered for choosing the next attribute to be added to the solution. For instance, instead of inserting the nearest city, one can compute for each possible city if selected what would be its next nearest city and add these two amounts together. The city with the smallest amount is then chosen.

Improvement Phase of the BDM (Step 2)

The obtained solution in Step 1 can be tested for possible improvement by simply trying to check whether a better solution can be found in the neighbourhood of the current solution, usually referred to as the incumbent solution. The neighbourhood structures and the selection mechanism adopted are crucial elements that contribute considerably to the success of any heuristic including BDM. Some of the neighbourhood structures are defined earlier. The commonly used selection strategies are given here.

Selection Strategies

In Step 2 of the BDM (or in any improvement or local search step of a given heuristic), the choice of the selected solution $X' \in N(X)$ can be based either on

(i) *the best improving move:*

— In the discrete case, evaluate all or a part of the neighbourhood and select the attribute that yields the least cost (case of minimisation problem):

$$X' = \text{ArgMin}\left\{ F\left(X''\right), X'' \in N(X) \right\}$$

— In the continuous case, we can use a direction-based method with a direction say $\vec{S}$ which can be gradient-based or just sample-based by evaluating a few points around the current solution and choosing the best.

$$X' = X + \alpha^* \vec{S} \quad \text{where} \quad \alpha^* = \text{ArgMin}\left\{ F(\alpha) = F\left(X + \alpha\vec{S}\right), \alpha \in R \right\}$$

(ii) *the first improving move:* Here, we select the first $X' \in N(X)$ that produces $F\left(X'\right) < F(X)$.

Note that (i) may take longer and can lead to a good local optimum and hence reduce the overall number of times the search is used. However, it can be hard to jump out of such a local minimum. In (ii), the selection is made relatively quicker and the first minimum may not be very attractive as it can be marginally better than the incumbent solution. In this case, the search may need to be repeated several times so to have the chance to explore other trajectories which may lead to even better local solutions than the one found in (i). Note that (i) has the advantage to accommodate the use of an appropriate data structure that will make it more efficient in the long run as will be shown in the implementation chapter (Chap. 6).

(iii) *a compromise strategy:* There is nevertheless another more flexible selection strategy that sits between (i) and (ii).

The idea here is not to choose the first improving solution as in (b), or wait until the entire neighbourhood is searched to choose the overall best, but to select the kth best improvement instead. This could be achieved, for instance, by accepting the move that yields an improvement that is considered to be good enough. This can be measured using a set threshold say 1 %, the best of a minimum number of improving moves (say 2 or 3), or just the best improving move found after a maximum time is elapsed since the previous improvement. Note that the latter is a time-related rule used to guarantee that the search does not go for a long period. One implementation would be to adopt these three rules whichever comes first. This strategy avoids the large amount of time required by (a) while controlling the solution quality by avoiding the drawbacks of (b). Note that (c) also does not have the flexibility to cater for simple data structures as will be discussed in Chap. 6, though more complex data structures could be attempted.

2.3   Classical Multi-Start

The BDM procedure is efficient when the objective function F is unimodular (i.e., F has one local minimum only). However, when F has several minima, it is impossible to get out of the neighbourhood of a local minimum using the same neighbourhood, see Fig. 2.1.

One way to increase the chance of getting better local minima is to restart the search randomly from another initial point (Step 1 in

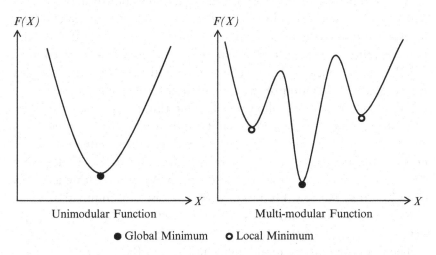

Fig. 2.1   Local optimality representation of a continuous function

Algorithm 2.1) in the expectation that one of them will lead to the right region which includes the global minimum. This is also commonly suggested in numerical optimisation software for the user to start from several initial points just in case the objective function happens to be non-convex. However, such a fully randomised multi-start mechanism can though better than nothing be considered as a blind search as it may lead to already visited local minima. There are, however, several multi-start schemes that are powerful enough to exploit the diversity of the search from one run to another, see Marti (2003), for instance. The idea of incorporating dissimilar solutions from one run to the other can be one way forward to provide diversity and hence guide the search.

2.4   GRASP

One possible compromise between BDM and the classical multi-start is to combine both randomness as in the classical multi-start and greediness as in BDM. Such an approach was initially introduced by Feo and Resende (1989), and then formally adopted by Feo and Resende (1995). This is known as greedy-randomised adaptive search procedure (GRASP). This is a multi-start heuristic where for each run, the search consists of two phases namely the construction phase (initial solution) and a local search phase for possible improvement. GRASP differs from multi-start in the way the initial solution is generated by combining randomness with greediness. GRASP can be considered as a memory-less multi-start heuristic. The idea is to use at each iteration of the construction of the solution a selection rule that is less rigid which allows flexibility in choosing not necessarily the best attribute but anyone in the vicinity of the best including itself. This greedy function (say $g(.)$) is used to evaluate the contribution of all the possible attributes and to define a restricted candidate list (RCL) using either the top k best attributes or those attributes that are within a certain deviation from the best. For instance, for the case of minimisation

RCL $=$ {set of attributes e such that $g_{min} \leq g(e) \leq g_{min} + \alpha(g_{max} - g_{min})$}; $0 \leq \alpha \leq 1$.

The attribute e^* is then randomly chosen from RCL.

Note that if $\alpha = 0$, this reduces to a greedy method (BDM), whereas if $\alpha = 1$, the search becomes the classical multi-start approach. For instance, when using the nearest neighbour heuristic for the TSP, instead of systematically choosing the nearest node from the current node (say i) as the next node for inclusion in the tour, a few nodes that are nearer to node i can be considered in the RCL from which one node will be chosen randomly or pseudo-randomly from this list. By allowing the selection rule to accommodate such flexibility while retaining the quality of the attribute to choose, this process has the power to obtain more than one solution compared to BDM besides not being totally random given it is controlled or guided by RCL. The main steps of a basic GRASP are given in Algorithm 2.2 but more details can be found in Resende and Ribiero (2010).

Algorithm 2.2: A Basic GRASP

Step 1   (Initial Phase) Select an initial solution, say $X \in S$ using a greedy-randomised procedure based on the RCL.

Step 2   (Improvement Phase/local search) Choose a solution $X' \in N(X)$ such that $F(X') < F(X)$. If there is no such X', X is considered as a local optimum with respect to this initial solution and go to Step 3, else set $X = X'$ and repeat Step 2.

Step 3   (Termination Phase) If $F(X) < F_{best}$ set $F_{best} = F(X)$ and $X_{best} = X$. If the maximum number of runs is performed stop, else go to Step 1.

Some Thoughts

There are several ways on how to explore further this approach. For instance, the choice of α does not need to be fixed or estimated beforehand but can be pseudo-randomly generated from one run to another depending on the solution quality found in previous solutions (this is known as reactive grasp). The simplest way is to run this approach for several values of α and then select those values that produced good quality solutions. These promising values could then be used as a subset from which to choose either uniformly or pseudo-randomly based on their respective solution quality. The definition of α can also be defined as a function of the gain/loss at a given iteration, or as a convex or a concave function in the number of iterations. A learning stage and additional guidance could also be embedded into the search to produce a better quality of the initial solution. For example, Luis et al. (2011) investigated some of these aspects with encouraging results when tested on a class of facility location problems. Though RCL provides some diversity, one could incorporate a diversity measure of any current solution before the local search step is activated on that solution, otherwise some changes to the solution could be made. This mini step may be worth introducing once in a while and after a minimum number of runs were applied so useful information would be recorded and acted upon in subsequent runs.

2.5   Simple Composite Heuristics

This sort of heuristic is similar in principle to BDM where in Step 2, one particular improvement procedure (also known formally as a local search) is used, whereas here the idea is to continue using one or two other improvement procedures in sequence, see Algorithm 2.3.

Algorithm 2.3: A Basic Composite Heuristic

Step 1   Initialise k possible refinements (usually $k = 2$ or 3) and generate an initial solution.

Step 2   Apply the k refinements in sequence on the incumbent solution.

Step 3   If there is improvement, repeat Step 2, otherwise record the last solution and stop.

Consider again the TSP for illustration. For instance, the original tour in Fig. 2.2a can be improved by exchanging the two arcs AB and CD with the two new arcs AC and BD resulting in a tour with a reduced length as shown in Fig. 2.2b. Note that the tour in Fig. 2.2b is a local optimal tour with respect to this 2-optimal procedure. I shall come back to this interesting property of local optimality later in this chapter.

It can be noted that using one single neighbouring search may not be sufficient enough to generate better quality solutions. For instance, the obtained tour given in Fig. 2.2b can be improved further using either

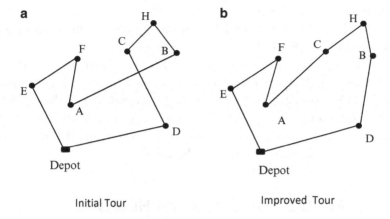

Fig. 2.2 A possible tour improvement

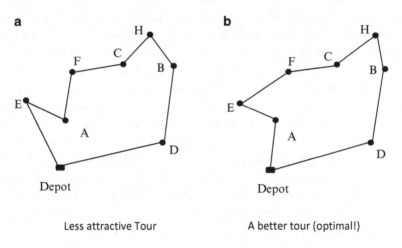

Fig. 2.3 More tour improvement

(a) another neighbourhood structure such as the 3-opt procedure. This can be achieved in one step by exchanging three existing arcs (OE, FA and AC) with three new arcs (OA, AE and FC) while retaining the structure of a tour, or (b) allowing a less attractive tour as the one shown in Fig. 2.3a. The optimal tour, for this particular example, is given in Fig. 2.3b. For instance, Salhi and Rand (1987) implemented, among other refinements, a simple version of the 3-opt procedure that considers successive arcs only for the case of the vehicle routing problem. Other enhanced implementations can also be found in the literature.

In brief, the r-opt procedure was initially developed by Lin (1965) which aims to exchange r arcs with r new arcs while maintaining the route structure. Obviously the larger the value of r the more likely there would be a larger improvement if it exists, but at the expense of a relatively much larger computational time. It was observed in the literature that the use of $r = 3$ is usually suitable for many applications.

2.6   A Multi-Level Composite Heuristic

In simple composite heuristics as described in Algorithm 2.3, the solution is improved using a small number of refinement procedures (local searches) put in sequence, and the method stops when there is no further improvement. Such a solution is then considered to be the best solution, which is a local minimum with respect to the selected procedures. *One way to direct the search out of a local minimum (maximum) is by using a large number of refinements (local searches) for which local optimality no longer holds. In addition, once a new better solution is found, it can be used as an initial solution for any of the other refinements, not necessarily the next one in the list. For simplicity, we can restart the process using the first one which is the simplest refinement in our list of available refinements.* This is the backbone of multi-level heuristics proposed in Salhi and Sari (1997) for solving the multi-depot routing with heterogeneous vehicle fleet. One possible skeleton of this approach with p levels is given in Algorithm 2.4.

The aim here is not to restrict the search to using two or three refinement procedures but to introduce as many as one wishes to as long as the total computing time remains acceptable and the overall approach is easy to understand. The choice of the refinement procedures, the sequence in which these are used and the choice of the refinement to go back to, once a new better solution is found, can be critical. For example, in Level 1, the solution can be generated as in the BDM. In Level 2, the composite heuristic of Type I which is the quickest may consist of several quick refinement modules that are implemented in sequence. These modules need to be fast.

This can be accomplished either because the procedures themselves are fast, or because they are applied to a relatively smaller and well-restricted neighbourhood. The next levels can be distinguished by using refinement modules that are more and more powerful and that require a larger and larger amount of computing time. These heuristics of a higher type (say Type III or IV) may be part of the refinements of the previous levels but applied to a larger neighbourhood. Note that once a solution is found at a given level, the solution is then changed in structure, and it is therefore appropriate to go to any of the other refinements of the earlier levels. A standard approach to revert back to the first level where computations are relatively fast is usually adopted. More details on this approach can be found in Salhi and Sari (1997). It is worth noting that this approach shares some similarities with variable neighbourhood descent which will be described next.

Algorithm 2.4: A Basic Multi-level Heuristic

Level 1   Generate an initial solution.
Level 2   Apply a composite heuristic of Type I.
Level 3   Apply a composite heuristic of Type II. If there is improvement, go to Level 2.

.
.
.

Level   Apply a composite heuristic of Type p-1. If there is improvement, go
p to Level 2.
Level   If none of the stopping criteria is met, perform a diversification
$p + 1$ mechanism and go back to Level 2, else record the final solution
            and stop.

Some Thoughts

The way the levels are organised can have an effect on the solution quality and hence some guidance along those lines may be useful. The local searches (the refinement procedures), some of which are based on the complete or partial enumeration of given neighbourhoods, may need to have a well-defined structure to make it easily accessible to a wider audience. The choice of successive refinements may be critical as one may wish to consider the next local search to be drastically different in structure to the previous one not necessarily the quickest so as to provide diversity where a new solution would be more likely to be found. The selection of the refinement where to go back to once a solution is determined does not need to be always Level 1 or even the same at each run. It may be that the obtained solution would be better suited to another refinement chosen randomly or based on certain rules that consider the structure of the current solution. This can be a challenging task to match the structure of a given solution with the characteristics of the set of refinements used. The integration of learning within the search could render this approach more adaptive and hence relatively more powerful.

2.7   Variable Neighbourhood Search

This type of heuristic attempts to escape local optimality by using systematically a new, usually larger, neighbourhood whenever there is no possible improvement through a local search in a given neighbourhood, and then reverts back to the first one, usually the smallest, if a better solution is found. Variable Neighbourhood Search (VNS) as originally developed by Mladenović and Hansen (1997) for solving combinatorial and global optimisation problems. VNS and its variants have been applied to several combinatorial problems with success; see Hansen et al. (2010). VNS, as with multi-level heuristics, takes advantage of the fact that a local minimum with respect to a given neighbourhood may not be a local minimum for other neighbourhoods. However, a global minimum is a local minimum for all the neighbourhoods. Another interesting fact which is reported in VNS but not shared in other metaheuristics is that local minima with respect to one or several neighbourhoods can be relatively close to each other. Compared to other powerful heuristics, VNS has the advantage of being simple and easy to implement as it requires only the definition of the neighbourhoods. Hence, there are relatively few sensitivity issues related to parameter calibration as in many other metaheuristics. The main steps of a basic VNS are given in Algorithm 2.5.

Variable Neighbourhood Descent This is a well-known variant of VNS where Step 2a and Step 2b are replaced by one step only using a descent type method based on the kth neighbourhood instead. In other words, the effect of randomness is reduced here. The neighbourhoods are explored either completely (the best improvement type strategy) or partially (first improvement type strategy), and the best solution X'' is then recorded. In other words, $X'' = \text{ArgMin}\{F(X'), X' \in N_k(X)\}$. All other steps of Algorithm 2.5 remain unchanged.

Algorithm 2.5: A Basic VNS Heuristic

Step 1   Generate an initial solution and select the set of neighbourhoods that will be used, say N_k, $k = 1,\ldots,k_{max}$ where k_{max} denotes the maximum number of neighbourhoods. Start with $k = 1$ (i.e., the first neighbourhood).

Step 2   While none of the stopping criteria (such as the total computing time, t_{max}) are met, perform the following tasks:

(i) Generate a point X' randomly in the neighbourhood $N_k(X)$.
(ii) Apply a local search starting with X' and let X'' be the obtained solution.
(iii) If the new solution X'' is better than X, set $X = X''$ and $k = 1$
    Else if $k = k_{max}$ set $k = 1$ (i.e., return to N_1), else set $k = k + 1$
(iv) Return to the top of Step 2.

Reduced VNS This acts as the opposite of variable neighbourhood descent (VND) where Step 2b is made void (i.e., no local search is activated). This simpler version is efficient when the problem is too large for a local search to be used frequently as this can be too expensive. In other words, a shake is used continuously to get a good random point within the neighbourhood that will be used as incumbent and the process is repeated. As an optional step, a local search could also be added here but just once, at the end of the search, turning reduced VNS (RVNS) into a composite heuristic.

General VNS Note that in Step 2(ii), the local search can be made up of a composite heuristic or even of a powerful metaheuristic though considerations on the computational effort need to be balanced. In general VNS (GVNS), to remain within the VNS framework, VND is used in Step 2(ii).

Skewed VNS The aim of this variant is to provide flexibility in selecting the new incumbent even if this X'' found in Step 2(ii) is not better than the best but not too far away from it. This aspect is similar to the use of threshold accepting which I shall discuss in the next chapter. In other words, X'' could be chosen

$$\text{if} F(X) < F\left(X''\right) \leq F(X) + \Delta_X^{X''} \text{ with } \Delta_X^{X''} > 0 \text{ being defined as follows:}$$

$\Delta_X^{X''} = \alpha.d\left(X, X''\right)$ or $\Delta_X^{X''} = \beta\left[\left(F\left(X''\right) - F(X)\right)/F(X)\right]$ with α and β used as control parameters. In other words, we revert back to $k = 1$ in Step 2(i) instead of exploring a larger neighbourhood ($k = k + 1$) while obviously recording the overall best solution. Note that this variant can fit into the next category where non-improving moves are also accepted (Chap. 3), but for simplicity and continuity in VNS related variants, this is kept here. This variant is powerful but uses an additional control parameter (α or β), which may go against the basic philosophy and simplicity of VNS.

Decomposition VNDS When the problem is large, another way would be to decompose the problem into smaller problems and solve each sub-problem by VNS individually while retaining the already obtained values of the variables of the other sub-problems.

Formulation Solution Space VNS This is an interesting variant where different formulation spaces for the same problem are used as proxy for the neighbourhood structures. For instance, the same problem can have two formulations, namely, F_1 and F_2. The search consists of two stages where in Stage 1 the search based on F_1 starts using any suitable heuristic technique or one of the VNS variants until the best local minimum is found. This solution is then transformed into F_2 and its corresponding configuration is used as a starting solution in Stage 2 leading to a new best solution which is then fed back to Stage 1 and the process continues until there is no improvement in either stages. The process could also start using F_2 first then F_1 instead. For example, Mladenović et al. (2005) propose this novel mechanism of alternating between two formulations for solving the circle packing problem with exciting results.

Two Level and Nested VNS These two recent and related VNS variants use at Level 1 VNS an inner level to carry out the intensification in depth within an outer level known as Level 2 VNS. This two level approach was initiated by Mladenović et al. (2014) and the nested approach by Brimberg et al. (2015). Both approaches share some similarities with the nested heuristic approach developed for the location-routing problem by Nagy and Salhi (1996). This VNS approach is efficiently adapted by Wassan et al. (2016) for a class of routing problems with backhauling. The depth of the local search engine at each level is crucial and needs to be carefully exploited.

For more information and variants of VNS, the reader will find the survey paper by Hansen et al. (2010) to be interesting and informative.

Some Thoughts

Possible extensions could include relaxing slightly Step 2(i) by generating a few solutions, say K, instead and take the best, or generating a few solutions till a certain threshold from the best is attained. These solutions could be generated in a guided manner or even randomly within that given neighbourhood. Imran (2008) in his PhD dissertation explored this aspect with mixed results when tested on a class of routing problems.

The choice of the next neighbourhood to evaluate could also be based on a selection criterion that relies on an approximation function evaluation or lower bounds. The return to the first neighbourhood could be relaxed to include the most appropriate one among the earlier neighbourhoods. The latter needs extra care as the additional computational burden of this extra task has to be weighed against the benefit and the simplicity in choosing the neighbourhoods systematically.

In Step 2(ii), one way would be to have the local search engine made up of several local searches where a subset of these local searches is selected at each iteration either randomly or pseudo-randomly leading to the so called *Random VNS* or adaptively turning it into an *Adaptive VNS*.

The idea of alternating between the formulations as attempted in formulation solution space (FSS) instead of shifting between the neighbourhoods is an exciting research avenue. The issue is that the problem needs to have at least two formulations. A similar and interesting approach recently developed by Brimberg et al. (2014) which can have a lot of scope is to perform the shifting step between equivalent problems, namely, a discrete location problem and its counterpart the continuous one instead of formulations.

2.8 Problem Perturbation Heuristics

The idea is to perturb the original problem to obtain a series of gradually perturbed problems. In other words, the ith problem has a slight perturbation or violation from the $(i-1)^{th}$ problem. Starting from a feasible solution to the problem found heuristically, we add a violation to define the next perturbed problem which we then solve heuristically. As there is a well-defined perturbation built up between successive perturbed problems, the successive solutions may have the tendency to retain some of the important attributes in their respective configurations. Once we reach the most perturbed problem, we can then start relaxing the problem by gradually removing the violation from one problem to another. This reversal process is carried out until we reach the feasible problem again. At this stage, we can either start adding violations again or making the original problem even less restricted by removing some of the constraints one at a time to form a new relaxed problem that is solved. Once we obtain the most relaxed problem, we can then start adding constraints till we reach our original problem from which we then continue adding violations. This process is repeated several times until there is no improvement in the original problem.

For instance, an investigation that adopts the above approach is conducted for the p-median problem where the objective is to identify the optimal location of a fixed number of depots (say p) with the aim to minimise the total transportation cost. Here, the solution is allowed purposely to become infeasible in terms of the number of depots. This is done by accepting solutions with more than or less than the number of depots required by the problem (say $p \pm q$ where $q \in [p/4; p/3]$). By solving the modified problems, infeasible solutions can be generated that when transformed into feasible ones may yield a cost improvement. Initially, a feasible solution with p facilities is found, then one depot is added at a time where a low-level local search (LS_1) is activated. Once we reach $p + q$ facilities, we then start dropping one depot at a time till we get to a feasible solution with p facilities where a higher level and a more powerful local search (LS_2) is applied for intensification purposes to try to obtain a better local minimum. At this stage, we also allow the solution to be infeasible by accepting less p facilities than required by removing one facility at a time until we reach say $p - q$, from where we start adding again facilities until we reach a feasible solution again. This up and down trajectory makes up one full cycle. The process is then repeated until there is either no improvement after a certain number of cycles or the overall computing time is met. Note that during the search process, the number of open depots can go down to $(a)\, p - q$ or up to $(b)\, p + q$. In (a) the number of depots is small enough to nearly guarantee that those depots will remain in the final solution as these must have strong and useful features, whereas in (b), the solution has a relatively large number of depots, most of which are competing depots. When this process is repeated several times going through several cycles, it acts as a *filtering process* where the most attractive depots will have the tendency to remain in the best depot configuration (survival of the fittest). Such an approach was developed by Salhi (1997) and it performed rather well when tested on a class of large facility location problems with known and unknown values of p. Zainuddin and Salhi (2007) adapted this approach to solve the capacitated multisource Weber problem.

Related Approaches

The idea of shifting between feasible and infeasible solutions is also used in strategic oscillation as one of the ways of tackling the issue of constraint handling which I shall briefly review in the next chapter under the tabu search section and also revisit in the implementation chapter. Here, the amount of infeasibility allowed is dealt with by attaching penalties in the objective function that are dynamically updated. This concept is useful particularly when the solution space is disconnected or non-convex as it is impossible to cross the infeasible region using only feasible solutions. Note that this shortcoming could become less severe if larger jumps are incorporated into the search as in large neighbourhood search (LNS), for example. The only difference with problem perturbation heuristics is that the latter is designed purposely to explore these infeasible solutions in a guided manner without any penalty attached so that some useful and promising attributes could then be identified. A useful application in solving a class of 0–1 multi-dimensional knapsack problems was also conducted successfully by Hanafi and Freville (1998) where their tabu search shifts between feasible and infeasible regions in a guided manner.

Another approach that aims to perturb the solution instead of the problem is conducted by Schrimpf et al. (2000) for solving routing problems, which they refer to as the 'ruin and recreate' procedure. This will be discussed under LNS in the next section.

A related method that perturbs the problem space by introducing random noises to the data is the noising method developed by Charon and Hudry (1993). The idea is to solve the problem with the perturbed objective function values via a local search where at each iteration, or after a fixed number of iterations, the level of perturbation is reduced until it reaches zero and the problem reverts to the original one. The way the noises are introduced and how randomness is gradually reduced as the search progresses are two key factors that are critical to the success of the method. Relatively recent work by Charon and Hudry (2009) explored the self-tuning of the parameters of such a method.

Some Thoughts

The way the perturbation is applied is vital as some guidance in the search to return to feasible solutions is necessary. If no control is embedded into the search, it may go astray and it may therefore become difficult to revisit promising regions or to even revert to feasible areas. For instance, in the p-median problem, the control is performed through a gradual increase or decrease in the number of facilities. Note that in the case when the number of depots is not known, the method can easily cater for such a change by considering the number of facilities found for the best current feasible solution as if it is p which obviously keeps being updated if a new best solution is found. The choice of q does not have to be fixed beforehand and the intermediate jumps do not have to be incremented by one but can be made to change dynamically as the search progresses. Very recently, an interesting study that considers this dynamic effect using learning is explored by Elshaikh et al. (2016) who produced competitive results when tested on the $p-$ centre problem on the plane.

2.9   Other Improving Only Methods

Large Neighbourhood Search

The idea is to use not necessarily small neighbourhoods but larger ones as well. The need for these large neighbourhoods is sometimes crucial to get out of local optimality as pointed out by Ahuja et al. (2002). This was proposed by Shaw (1998) and can be seen to be similar to the 'ruin and recreate' procedure of Schrimpf et al. (2000).

The idea is to perturb the solution configuration in an intelligent way by deleting some of the attributes of the solution using some removal strategies and then reintroducing them using some insertion strategies. The main aspects include the number of attributes to remove (K), the removal strategies adopted and the insertion strategies used. The way these strategies are implemented and developed is critical to the success of the search. As the method is repeated several times, some form of learning is worth exploring to efficiently guide the search.

Choice of K The value of K has a direct effect on the perturbation as a large value will destroy the main structure of the solution which can be inefficient, whereas a small value may have little bearing on the changes. However, the above claim could be taken further by making K to be a discontinuous and non-increasing function in terms of the number of iterations (*iter*) in the range $[K_{min}, K_{max}]$ as follows:

$$K = \begin{cases} K_{max} & \text{if} & iter \leq T_{min} \\ g(iter) & \text{if} & T_{min} \leq iter \leq T_{max} \\ K_{max} & \text{if} & iter \geq T_{max} \end{cases}$$

For instance, at the initial stage of the search for a certain number of iterations (say T_{min}), we may encourage strong perturbations using $K = K_{max}$ which may produce diverse and good solutions. However, at the end of the search, after a certain number of iterations were performed (say T_{max}), we aim to inject small changes only to reach our overall best solution by setting $K = K_{min}$. For the rest of the computations, we could define $K = g(iter)$ to be a non-increasing function (linear with negative slope, or negative exponential, etc.). Another way would be to find the value of K pseudo-randomly in $[K_{min}, K_{max}]$ where in the learning phase, K is generated randomly in the range and a score related to each value is recorded.

Removal Strategies These include (a) the pure random removal strategy where K attributes are randomly removed from the solution configuration. For the case of the TSP, say K nodes are deleted from the tour, (b) informed removal strategies that aim to use some selection rule for removing the K attributes. For the TSP, this may include the K nodes with longest edges in the tour, or those with small customer demand to insertion cost ratios, among others. At a given iteration, a given removal strategy can be chosen randomly or based on a pseudo-random choice which can be defined beforehand or found through learning. One way would be to adhere to a two-stage process where in Stage One, a learning phase is performed where each strategy is chosen either randomly or applied for the same number of times while recording its success score. In Stage Two, this information is then used to pseudo-randomly select at each iteration the removal strategy.

Insertion or Repair Strategies These K removed attributes need to be reinserted back either one by one, in smaller subsets (small chains), or optimally assigned into their best position. This can be conducted in a greedy way using the classical 'least cost insertion' rule or in a flexible manner as adopted in the construction phase of GRASP, or using an optimal assignment method. Note that the latter could be expensive as many runs are usually required.

Note that LNS and problem perturbation methods share some similarities in the way they rely on discovering interesting features through ruin and repair. However, LNS focuses on perturbing the solution in one step, whereas the other concentrates instead on perturbing the problem gradually.

Iterated Local Search

This consists of two main phases known as the construction/diversification phase and the local search phase. These are performed repeatedly while an acceptance step is embedded to retain the best solution found so far. The aim is to ideally keep using the same local search as a black box while providing interesting solutions to it through diversification. There are several ways on how to implement this approach.

(i) The simplest one is to have a random solution every time making the approach resembles the classical multi-start but as we mentioned earlier, this approach could easily lead to poor solutions.

(ii) Another way would be to apply a randomised construction heuristic instead in Phase One where at each iteration, a new solution will be generated similar to the construction phase of GRASP. This obviously provides diverse initial solutions every time this construction is applied while retaining the quality of the solution. It is worth mentioning that such a randomised mechanism can be incorporated in all selection rules based on greediness to provide extra flexibility to the search and hence enhance the quality of the solutions.

(iii) The most interesting and challenging way, which is also commonly used, is to apply at each iteration a diversification scheme on the solution obtained by the local search. The diversification can be achieved by perturbing the best solution so far as in LNS, for instance, or through a mechanism that takes into account the history of the previously found solutions. The acceptance rule is usually to retain the overall best solution when reverting back to the diversification phase or using a softer acceptance rule that allows some level of deterioration.

Note also that the local search engine could be made up of several local searches. Though this can be powerful it may require excessive computational effort. Such a black box could then, for efficiency reasons, incorporate a randomised smaller set of local searches chosen from the entire set of local search operators either randomly or pseudo-randomly at each iteration. This approach is simple and efficient and shown to be powerful in obtaining interesting results in several combinatorial problems. One possible implementation of a strong ILS could be the combination of LNS (diversification/construction) with a VND or randomised VND (local search). For more details and possible applications on this heuristic, see Lourenco et al. (2010).

Guided Local Search

This is an adaptive local search which attempts to avoid local optimality and guide the search by <u>using a modified objective function</u>. This objective function contains the original objective and a penalty term which relates to the less attractive features of the configuration of a given local optimum. The local search used can be a simple improvement procedure or a powerful heuristic. In other words, at every local solution, the feature which has a large utility value receives an increase in its unit penalty. For instance, in the case of the TSP, the edges can represent features of the tour and the largest edge of the obtained tour (say edge, e) will have its penalty increased (say $p(e) = p(e) + 1$). Note that this type of penalty is related to the so called 'bad' features and not to the amount of infeasibility as usually carried out in constrained optimisation. The way the features are selected and penalised play an important part in the success of this approach. Voudouris and Tsang (2010) provide an interesting review with an emphasis on how to implement this approach when addressing several combinatorial problems.

2.10 Summary

Some of the heuristics that are based on improving moves only are described in this chapter, some with more depth than others. The idea of moving from one solution to a better one is challenging. However, the design of suitable neighbourhoods, alongside efficient local searches could make the search more resilient and flexible enough to find good quality local minima if these exit. In brief, the approaches aim to achieve a better local minimum by systematically using various neighbourhoods as in VNS, applying strong local searches as in multi-level heuristic or VND, perturb the solution and repair it as in LNS, or perturb the problem and solve it then gradually get to the solution, or a combination of these through ILS, for example. The next chapter will deal with the case where inferior solutions are also considered within the search.

References

Ahuja, R. K., Ergun, O., Orlin, J. B., & Punnen, A. P. (2002). A survey of very large scale neighbourhood search techniques. *Discrete Applied Mathematics, 123*, 75–102.

Brimberg, J., Drezner, Z., Mladenović, N., & Salhi, S. (2014). A new local search for continuous location problems. *European Journal of Operational Research, 232*, 256–265.

Brimberg, J., Mladenović, N., Todosijević, R., & Urosević, D. (2015). Nested variable neighbourhood search. *SYM-OP-IS, XLII Int Conf on Oper Res,* September16–19, Ivanjica.

Charon, I., & Hudry, O. (1993). The noising method- a new method for combinatorial optimization. *Operations Research Letters, 14,* 133–137.

Charon, I., & Hudry, O. (2009). Self-tuning of the noising method. *Optimzation, 58,* 1–21.

Elshaikh, A., Salhi, S., Brimberg, J., Mladenović, N., Callaghan, B., & Nagy, G. (2016). An Adaptive perturbation-based heuristic: An application to the continuous p-centre problem. *Computers and Operations Research, 75,* 1–11.

Feo, T. A., & Resende, M. G. C. (1989). A probablistic heuristic for a computationally difficult set covering problem. *Operations Research Letters, 8,* 67–71.

Feo, T. A., & Resende, M. G. C. (1995). Greedy randomized adaptive search procedures. *Journal of Global Optimization, 6,* 109–133.

Hanafi, S., & Freville, A. (1998). An efficient tabu search approach for the 0–1 multidimensional knapsack problem. *European Journal of Operational Research, 106,* 659–675.

Hansen, P., Mladenović, N., Brimberg, J., & Moreno Perez, J. A. (2010). Variable neighbourhood search. In M. Gendreau & J. Y. Potvin (Eds.), *Handbook of metaheuristics* (pp. 61–86). London: Springer.

Imran, A. (2008). An adaptation of metaheuristics for the single and the multiple depots heterogeneous fleet vehicle routing problems. PhD thesis, University of Kent, Canterbury.

Lin, S. (1965). Computer solutions of the travelling salesman problem. *Bell Systems Technical Journal, 44,* 2244–2269.

Lourenco, H. R., Martin, O. C., & Stutzle, T. (2010). Iterated local search: Framework and applications. In M. Gendreau & J. Y. Potvin (Eds.), *Handbook of metaheuristics* (pp. 363–397). London: Springer.

Luis, M., Salhi, S., & Gabor, N. (2011). A guided reactive GRASP for the capacitated multi-source Weber problem. *Computers and Operations Research, 38,* 1014–1024.

Marti, R. (2003). Multi-start methods. In F. Glover & G. A. Kochenberger (Eds.), *Handbook of metaheuristics* (pp. 355–368). London: Kluwer.

Mladenović, N., & Hansen, P. (1997). Variable neighbourhood search. *Computers and Operations Research, 24,* 1097–1100.

Mladenović, N., Plastria, F., & Urosević, D. (2005). Reformulation descent applied to circle packing problems. *Computers and Operations Research, 32,* 2419–2434.

Mladenović, N., Todosijević, R., & Urosević, D. (2014). Two level general variable neighbourhood search for attractive travelling salesman problem. *Computers and Operations Research, 52,* 341–348.

Nagy, G., & Salhi, S. (1996). Nested location routing heuristic using route length approximation. *Studies in Locational Analysis, 8,* 3–22.

Resende, M. G. C., & Ribiero, C. G. (2010). Greedy randomized adaptive search procedures: Advances, hybridisations, and applications. In M. Gendreau & J. Y. Potvin (Eds.), *Handbook of metaheuristics* (pp. 283–319). London: Springer.

Salhi, S. (1997). A perturbation heuristic for a class of location problem. *The Journal of the Operational Research Society, 48,* 1233–1240.

Salhi, S., & Rand, G. K. (1987). Improvements to vehicle routing heuristics. *The Journal of the Operational Research Society, 38,* 293–295.

Salhi, S., & Sari, M. (1997). A Multi-level composite heuristic for the multi-depot vehicle fleet mix problem. *European Journal of Operational Research, 103,* 78–95.

Schrimpf, G., Schneider, J., Stamm-Wilbrabdt, H., & Dueck, H. (2000). Record breaking optimization results- using the ruin and recreate principle. *Journal of Computational Physics, 159,* 139–171.

Shaw, P. (1998). Using constraint programming and local search methods to solve vehicle routing problem. In CP-98 (fourth international conference in principles and practice of constraints programming). *Lecture Notes in computer Science, 1520,* 417–431.

Voudouris, C., & Tsang, E. P. K. (2010). Guided local search. In M. Gendreau & J. Y. Potvin (Eds.), *Handbook of metaheuristcs* (pp. 321–361). London: Springer.

Wassan, N., Wassan, N.A., Nagy, G., & Salhi, S. (2016). The Multiple trip vehicle routing problem with backhauls: Formulation and a two-level variable neighbourhood search. *Computers and Operations Research.* doi:10.1016/j.cor. 2015.12.07.

Zainuddin, Z. M., & Salhi, S. (2007). A perturbation-based heuristic for the capacitated multisource Weber problem. *European Journal of Operational Research, 179,* 1194–1207.

3

Not Necessary Improving Heuristics

3.1 Simulated Annealing

The concept of simulated annealing (SA) is derived from statistical mechanics which investigates the behaviour of very large systems of interacting components such as atoms in a fluid in thermal equilibrium, at a finite temperature. The question is what happens to the system at its low-energy ground state. Do the atoms remain a fluid, or do they solidify? Is this a solid crystalline? The way to lower the temperature is crucial to obtain a good crystal or glass. The system must be melted and then carefully cooled. This is referred to as an annealing process. In this process, a high temperature is first used to melt the solid and then gradually cooled, spending sufficient time at each time for the solid to reach thermal equilibrium, especially when approaching the freezing point. If the process of cooling is too rapid, then defects can be frozen into the solid, and the ground state will not be reached. The resulting solid may be only a metastable, or in combinatorial optimisation language, a locally optimal system, a structurally defective crystal or glass. In 1953 Metropolis et al. produced an exciting algorithm to provide efficient simulation of a collection of atoms in equilibrium at a given temperature. At each

© The Author(s) 2017
S. Salhi, *Heuristic Search*, DOI 10.1007/978-3-319-49355-8_3

iteration, an atom is given a small random displacement and the resulting change in the energy of the system is evaluated. If the change is negative, the move is systematically accepted as it is an improving move and the process continues using the new configuration as its starting configuration. However, if the change is positive, though this is a deterioration of the solution, such a solution is not systematically rejected but the new solution could be accepted if it passes a certain probabilistic threshold. In other words, the new solution may not be too far away from the best and therefore it may be worth considering it. This selection process will be defined more explicitly next.

The Physical Analogy

It was not until 30 years after Metropolis et al. (1953) study that the analogy between the simulation of the annealing of solids and the solution of large combinatorial optimisation problems was first made by Kirkpatrick et al. (1983). In the context of an optimisation problem, the process of a body cooling is an analogy for a search algorithm to seek a good solution with the aim to approach the global optimum. The energy function in our case will represent the objective function of the problem. The configuration of the system's particles becomes the solution configuration of the problem (i.e., parameter values, sequence in a tour, etc.). The search for a low-energy configuration is equivalent to the search for optimal solutions. The temperature will act as the control parameter. A hasty cooling resulting in defective crystal is analogous to a neighbourhood search that yields a poor local optimum. Kirkpatrick et al. adopted these analogies for the context of combinatorial optimisation, to explicitly formulate the algorithm that is now widely known as SA which is described in Algorithm 3.1.

The way in which the temperature decreases is called the cooling schedule and the probability function used is that of Boltzmann's Law (see Metropolis et al. 1953) which is a negative exponential function. The reasoning behind this choice is that it has the tendency to choose more non-improving solutions initially but as the search progresses this random-based technique will have a smaller probability of selecting inferior solutions due to their acceptance rule based on the temperature. For more details, see Eglese (1990), Osman and Laporte (1996), Drezner and Salhi (2002) and Dowsland and Thompson (2012).

Algorithm 3.1: A Basic SA Heuristic

Step 1   Select an initial solution X and set the iteration counter to $k = 0$. Choose the initial and final temperature values T_0 and T_f respectively, M the maximum number of iterations within the inner loop, and $iter_{max}$ the overall maximum number of iterations. Initialise $X_{best} = X$ and $F_{best} = F(X)$.

Step 2   Choose a solution $X' \in N(X)$ and compute $\Delta = F(X') - F(X)$.

Step 3   Do the following step M times. If $\Delta \leq 0$ or $\Delta > 0$ but $e^{-\frac{\Delta}{T_k}} \geq \theta$ with $\theta \in [0, 1]$, then accept the new solution X' and set $X = X'$. If $F(X') < F_{best}$ then set $F_{best} = F(X')$ and $X_{best} = X'$, otherwise keep X.

Step 4   If some stopping criteria are satisfied, take X_{best} as your best solution and stop, else update the temperature $T_{k+1} = g(T_k) \leq T_k$; set $k = k + 1$ and go to Step 2.

A Simple Example of a Basic SA Implementation

Consider for simplicity the minimisation of the following real valued one-variable function in the range $[-4, +6]$:

$$F(X) = \frac{x^6}{6} - \frac{2x^5}{5} - \frac{13x^4}{5} + \frac{14x^3}{3} + 12x^2$$

The optimal solution can be found analytically at $x^* = 4$ which yields $F^* = -68.266$. In this example, we illustrate how the starting solution can yield a local optimal solution which is rather poor compared to the global solution if descent methods were used instead. We assume we do not apply differentiation as in practice solving, $F'(.)$ could be a tedious task to handle.

A simple implementation of SA using the following cooling schedules is adopted.

– Initial temperature T_0 and M (Step 1)

   $T_0 = 1000$ (high temperature; uphill moves are likely to be all accepted) and for simplicity, let $M = 1$

– Updating of the temperature T_k (step 4)

   (a) $T_{k+1} = T_k - t$ (t = constant, say 5)
   (b) $T_{k+1} = T_k - t_k$ (t_k is randomly chosen in $[0, 5]$ at each iteration k)

– Stopping criteria (Step 4)

   $T_f \leq 5$ or the maximum number of iterations is 3000, whichever is reached first.

The performance of this basic SA heuristic is compared against a DM using initial solutions starting from -50 to 10 in a step size of 5. The results are summarised in Table 3.1

According to the obtained results, it can be rather misleading if we use DM starting from anywhere with a value less than 2. Three local optima are discovered, from which only one is the global one. This demonstrates the weakness of DM as in practice, the function can be too complicated and identifying interesting starting points is itself a complex problem. As we can see, SA has the flexibility of going over the hill to reach the global solution from all the initial solutions we started from. It is, however,

Table 3.1 The effect of the initial solution when using both SA and DM

Initial solution	SA: case (a) $x^*[F(.)]$	SA: case (b) $x^*[F(.)]$	Descent method (DM) $x^*[F(.)]$
−50	−2.64 [−20.805]	−2.64 [20.805]	−2.64 [−20.805]
//	//	//	//
−5	//	//	//
0	//	//	0 [0]
5	//	//	3.07 [47.622]
10	//	//	//

worth noting that the SA will obviously require more computational time than DM.

The cooling schedules used in this example are simple ones. In the next subsection, I present some of the commonly used ones.

SA Key Elements

The success of the SA heuristic is determined by the choice of the neighbourhood as well as the cooling schedules. The latter is defined by a number of different parameters including T_0, T_f, T_k, M_k, the stopping criteria, the neighbourhood structure and the objective function evaluation. The choice of the parameters is obviously not unique as this may depend on the characteristics of the problem. However, the following cooling schedules have shown to be promising when tested on a variety of combinatorial problems.

Initial Value of the Temperature

The value of the initial temperature (T_0) should be large enough to accept a large number of moves. In other words, $e^{-\Delta/T_k} \approx 1$ for nearly all the moves generated. However, if the value of T_0 is too large, most solutions will be initially accepted which can be a waste of time as many solutions should not really be considered. On the other hand, if the value is relatively small, many non-improving solutions will be rejected and this falls into the local optimality trap of the basic greedy method.

One possible computation of T_0 is by Aarts and van Laarhoven (1985)

$$T_0 = \frac{\overline{\Delta}}{A} \qquad \text{with} \qquad A = \text{Log}\left[\frac{m^+}{\rho m^+ + (1-\rho)(m - m^+)}\right] \qquad \text{with}$$

$\rho \in [0.5, 1]$; say $\rho = 0.8$
where

m represents the number of moves first generated in the neighbourhood of the initial solution $X, N(X)$,

m^+ the number of moves that produced a cost increase,

$\overline{\Delta}$ the average cost increase over the m moves and

$$\Delta_{\text{Max}} = \text{Max}\left\{ \left| F(X') - F(X'') \right| ; X', X'' \in N(X) \right\}$$

Van Laarhoven and Aarts (1987) simplified the above rule by relating T_0 to $\overline{\Delta}$ and Δ_{max} as follows:

$$T_0 = \frac{-\Delta_{\text{Max}}}{Ln(\rho)} \qquad \text{or} \qquad T_0 = \frac{-\overline{\Delta}}{Ln(\rho)} \qquad \text{or} \qquad \text{simply}$$

$T_0 = \lambda\overline{\Delta} + (1 - \lambda)\Delta_{\text{Max}}$ with $\lambda \in [0.25; 0.50]$

The Final Temperature T_f

The value of T_f needs to be small enough so that only improving solutions would be accepted, making SA acting as a standard hill climbing method at the very end. For instance, T_f can be fixed a priory to a small value, or made dependent on an acceptance threshold, say $\varepsilon \simeq 0.05$. This is equivalent to having $e^{-\frac{\Delta}{T_k}} \le \varepsilon$ leading to $T_f \le \frac{-\Delta}{\text{Log}(\varepsilon)}$. The value of T_f can also act as a stopping criterion.

The Updating of the Temperature

The temperature T_k theoretically needs to be a non-increasing function of iteration k (e.g., $T_{k+1} = g(T_k) \le T_k$). The possibility to allow the temperature to remain constant or to increase marginally at a given iteration is a flexible and effective option.

(i) *Constant step size:* $T_{k+1} = T_k - t$ where t a constant step size that could depend on the number of iterations, as well as T_0 and T_f. One way would be to express $t = m.\frac{T_0 - T_f}{M_k}$ where m and M_k represent the maximum total number of runs and the number of runs within each value of T_k (inner loop of the SA), respectively.

(ii) *Decreasing step size:* $T_{k+1} = \beta_k T_k$ where $\beta_k \in [0.80; 0.95]$ can be randomly generated once or at each iteration. This is a commonly used cooling schedule which is simple and easy to use besides shown to be effective when applied in several combinatorial problems.

(iii) *Range dependent:* $T_k \in [A_k, B_k]$ which can be disjoint ranges or overlapping ones.

Flexibility in Resetting the Temperature

The idea to reset the temperature at higher values after getting stuck in a flat region, say if no improvement is found after a certain number of iterations (i.e., consecutive uphill rejections). The current solution is locally optimum and since the temperature is low, the SA algorithm becomes self-destructive as it restricts the acceptance of less-attractive solutions. These non-improving moves seem to be rejected with probability nearly one. One idea to override such a limitation is to accept uphill moves by resetting the temperature. The search will have more chance to accept some solutions which could then provide a way to get out of the flat region of this current local optimum. The following resets are commonly used based on the temperature when the best solution was found, say T_{best} and the last temperature reset T_R.

$T_{k+1} = T_{\text{best}}$, or $\alpha T_{\text{best}} + (1 - \alpha)T_k$ or $\beta T_{\text{best}} + (1 - \beta)T_R$ (initially $T_R = T_{\text{best}}$)

with $\alpha, \beta \in [0, 1]$ representing the corresponding weight factors.

Some of these resetting schemes are initially given by Conolly (1990) and successfully applied by Osman and Christofides (1994) to solve the vehicle routing problem (VRP).

One may argue that the update function using such a reset scheme violate the property of $g(T_k)$ as it should be theoretically a non-increasing function of k. This is true if $g(T_k)$ was optimally defined but not heuristically as derived here. Such a statement needs, therefore, to be relaxed and allowing such flexibility is appropriate. Another way of relaxing such an update would be to incorporate flexibility by allowing the temperature to increase relatively slowly with the change in the cost function whether accepted or rejected, see Dowsland (1993) and Dowsland and Thompson (1998).

The Evaluation Function Speed Ups

The computation of the acceptance test $e^{-\Delta/T_k} \geq \theta$ with $\theta \in [0, 1]$ contributes considerably to the computational burden as several evaluations are performed. One way would be to approximate $e^{\frac{-\Delta}{T_k}}$ at temperature T_k by its Taylor series of degree one namely $1 - \frac{\Delta}{T_k}$ instead. For instance, Johnson et al. (1989) conducted experiments on graph partitioning using such an approximation and observed that about 30 % reduction could be achieved without a significant difference in the solution quality. This interesting result shows the effect of integrating mathematical approximation theory in heuristic search design such as SA.

Another way is to relax the use of evaluating one neighbour at a time by randomly selecting $x^{'} \in N(x)$ not only once by r times instead leading to $\{x^{'}_1, \ldots, x^{'}_r\}$ and the best one $x^{'} = \text{ArgMin}\left\{C\left(x^{'}_i\right); i = 1, \ldots, r\right\}$ is then used for the testing. As the solution is already fine-tuned, the number of trials within the same temperature M_k could then be reduced accordingly.

Stopping Criteria

The following stopping criteria are usually used some of which are commonly adopted in other heuristics.

(i) A maximum total number of iterations or maximum computing time allowed
(ii) A minimal value of the final temperature, T_f
(iii) A number of resets since the last improved solution
(iv) A maximum number of function evaluations
(v) A maximum number of iterations since the last improved solution
(vi) Others

Some Thoughts

The success of SA is based on the choice of the neighbourhood and the cooling schedules especially the temperature update and the initial temperature which is also dependent on the quality of the initial solution. The idea of allowing the temperature to go down as well as up and to reset it if necessary has proved to be successful in vehicle routing, time tabling, clustering, generalised and quadratic assignment problems. The flexibility in generating a small number of neighbouring solutions K_{min} in Step 2 of the SA algorithm instead of one solution only is worth exploring. K_{min} could also be adaptively adjusted as the search progresses. The best of these solutions is then used in the selection process of Step 3. It could also be more appropriate to adopt other distributions in the sampling process rather than the uniform, and also not to restrict the search to using the Boltzmann distribution but some of its variants instead. The choice of one neighbourhood only may also be restrictive and using a few neighbourhoods instead by generating one neighbouring solution from each neighbourhood either randomly or pseudo-randomly may render SA more adaptive and hence more powerful. Though classical SAs do not incorporate local search, it could be interesting to embed this additional ingredient once in a while during the search, making the SA resembles a bit ILS. An initial and interesting theoretical study on the convergence of SA based on Markov chains is made by Lundy and Mees (1986), this aspect could be worth revisiting. For more information on SA, the reader would find the chapter by Dowsland and Thompson (2012) to be interesting.

3.2   TA Heuristics

This heuristic is a simplified version of SA. The idea is to avoid the probabilistic effect when selecting a non-improving move. This is achieved by allowing a deterministic worsening of the solution known as thresholding. In other words, if a solution happens to be just a tiny fraction worse (say 1 % or 2 %) than the current best solution, such a solution may be worth examining. Obviously, the choice of the value of this threshold does not have to be a constant throughout the search and the main question is therefore how to define it or update it. For instance, on one hand, if it is too large, poor solutions would be chosen which can make the search ineffective for returning to promising regions. On the other hand, if it is too small, the method will behave just like a descent type approach. Dueck and Scheuer (1990) were the first to propose formally such an approach and to present empirical studies against SA when solving the TSP. Another way is to record a list of the top non-improving solutions which can then be used to guide the threshold target (the size of such a list can be made constant or dynamically changing). The main steps of the TA heuristic are given in Algorithm 3.2. This is similar to SA except that Step 3 provides the flexibility in accepting a non-improving solution deterministically through thresholding instead.

Algorithm 3.2: A Basic TA Heuristic

Step 1   Select an initial solution X, set the iteration counter to k $= 0$ and the maximum number of iterations to $iter_{max}$. Choose an initial value for the threshold, $t_0 > 0$, and the updating function of t as $g(.)$ as a non-increasing function of $iter$ (the iteration count). Set $k = 0$, $X_{best} = X$ and $F_{best} = F(X)$.

Step 2   Find the local minimum $X' \in N(X)$ and compute $\Delta = F(X') - F(X)$. Set $k = k + 1$

(continued)

> **Algorithm 3.2** (continued)
>
> Step 3   If $\Delta \leq t_k$, accept the new solution X' and set $X = X'$. If $F(X') < F(X)$, set $X_{best} = X'$ and $F_{best} = F(X')$, else keep X.
>
> Step 4   If after a certain number of iterations there were too many acceptances, reduce the threshold t_k ($t_k = g(k, t_k)$) and go to Step 2.
>
> Step 5   If $k = iter_{max}$ record X_{best} and F_{best} as the solution and stop. Else go to Step 2.

TA has advantages over SA in the sense that TA is simpler to use, deterministic and has a smaller number of parameters. In addition, TA has shown to be computationally efficient when tackling hard combinatorial problems especially routing-based problems (see Tarantilis et al. 2003; Li et al. 2007). Besides, in Step 2, though for simplicity we refer to $N(X)$, the new solution is usually found through exploring the neighbourhood fully or partially acting as a local search leading to a local minimum. Due to this implementation, TA will require significantly a relatively smaller number of iterations than SA though in each iteration TA will be consuming much more computing time due to Step 2 computation. For further references on TA and its implementation, see Hu et al. (1995) and Lee et al. (2004).

The following two simple but successful variants of TA have proved to be promising and hence worth mentioning.

Record to Record Heuristic

This is developed by Dueck (1993) where the threshold is based on the best solution instead of the current solution as in the basic TA (i.e., Step 2 is modified to using $\Delta = \frac{\left(F(X') - F_{best}\right)}{F_{best}}$) instead and in Step 3, t_k represents the allowable percentage solution worsening. In other words, a solution that is worse by up to t_k% (say 10 % to start with) is accepted. This differs from the basic TA in the sense that (a) the approach takes into account the relative deviation from the best solution instead of the absolute deviation from the current solution, and (b) the reference point is the best solution instead of the current solution. In (a), the various magnitudes of the data in different instances could be misleading if the classical TA is blindly implemented, whereas the relative measure remains valid regardless. Li et al. (2007) adopted this approach for solving the heterogeneous vehicle routing problem.

List-Based TA

This approach is developed by Tarantilis et al. (2003) where instead of having a threshold value based on the objective function only, a list containing a number of the top solutions is used with its cardinality being the only parameter that needs to be controlled, and hence makes the search easier to implement. This is based on initially applying a local search and recording the top M solutions with their respective cost deviation from the best $\Delta_l = \frac{(F(X_l) - F_{best})}{F_{best}}$; $l = 1, \ldots, M$ to make the list of threshold values with $t_k = \text{Max}\{\Delta_l; l = 1, \ldots, M\}$ the largest threshold. In Step 2, a new selected solution $X^{'}$ is found using a local search, which may or may not be the same one used at the initial step. If $\Delta = \frac{\left(F\left(X^{'}\right) - F_{best}\right)}{F_{best}} \leq t_k$, then $X^{'}$ is selected, $X = X^{'}$ and the list is adjusted by dropping t_k, inserting Δ in the new list and recomputing t_k again. Otherwise X and t_k are both retained and the process continues. This mechanism produces a function of t_k which is a non-increasing function of k. During the search, the list is reduced gradually by decreasing the value of M and hence t_k. The authors produced competitive results when testing this scheme on a class of routing problem with heterogeneous vehicle fleet.

Some Thoughts

The success of TA relies on the way the threshold is updated namely the function $g(.)$.

It is worth noting that in both TA and SA, diversification strategies could also be embedded into the search say if no improvement has been observed after a certain number of successive iterations. This aspect is incorporated in the study of Li et al. (2007) with impressive results. As in SA, there are also many interesting variants that could be looked at. For instance, the linkage of the quality of the solution to the threshold could be dynamically adjusted and carefully designed. For instance, Hu et al. (1995) discussed one approach known as the old bachelor acceptance where the threshold is allowed to be moved up as well as down similar in principle to the idea of temperature resetting in SA.

3.3 Tabu Search (TS)

This approach was proposed by Glover (1986) and independently discussed by Hansen (1986) as a descent/ascent technique. TS concepts are derived from artificial intelligence where intelligent uses of memory help exploit useful historical information. In TS, attempts to overcome local optimality are made by accepting non-improving moves and <u>imposing some sort of tabu status</u> for those attributes recently involved in the move.

TS shares with SA and TA the flexibility in accepting non-improving moves. TS is one of the metaheuristics that has proved successful in solving several hard combinatorial optimisation problems. These include production scheduling, graph colouring and graph partitioning, telecommunication path assignment, vehicle routing, quadratic assignment, location problems and so on. Such an approach was also tested on an unusual optimisation problem, namely, the variable selection problem in linear regression by Drezner et al. (1999). The results were competitive when compared to the well-known statistical software SAS especially when the data are highly correlated.

I first give the basic TS terminology, an outline of the TS procedure followed by an explanatory of the main steps of the algorithm. The last subsection is devoted to some key components that make TS exceptionally powerful.

Basic Terminology

To be familiar with the TS vocabulary the following items are usually used.

A move: A transition from a current solution to its neighbouring (or another) solution.

An attribute: The elements that constitute the move; for instance, if the move is to add or drop customer $i(i = 1, \ldots, n)$ from a given route, the associated attribute is the ith customer, whereas if the move is formed by exchanging customer i with customer j between their respective routes, the attribute can be seen as the pair of customers i and j.

Tabu list: A list of moves that are currently made tabu.

Tabu list size: The number of iterations for which a recently accepted move is not allowed to be reversed (say $|T_s|$); this can also represent the number of tabu moves.

Tabu tenure: An integer number stating how long a given move will remain tabu (i.e., $0 \leq TAB(i) \leq |T_s|$).

Aspiration criterion: A threshold (usually the best current objective function value, or an estimate of that value, etc.) for which the tabu status of a move can be relaxed (i.e., making the tabu restriction void).

Admissible move: A move that is non-tabu or a move that is tabu active but which can produce a solution above the aspiration level.

Forbidding strategy: The tabu conditions that forbid a move from being reversed.

Freeing strategy: The conditions that allow a move to become non-tabu either because its tabu status has become not tabu or it satisfies the aspiration criterion.

The TS Algorithm

In brief, TS selects the next best move in a <u>deterministic manner</u> so it is aggressive as it exploits a larger part (or all) of the neighbourhood. TS tries to avoid cycling by stopping reversal moves for a certain number of iterations but it is flexible as it accepts tabu moves that have an outcome better than a certain defined aspiration level (i.e., this is usually the best objective function value). TS takes into account past information of already found solutions to construct short- and/or long-term memories. These memories are useful in guiding the search via diversification (the exploration of other regions) and intensification (going deeper and deeper within the same vicinity of the current solution which is achieved through the use of local search). TS, as other metaheuristics, has also the power of searching over non-feasible regions which, in some situations, can provide an efficient way for crossing the boundaries of feasibility. A basic TS procedure is given in Algorithm 3.3.

Note that the search over $N(X)$ acts as a local search as $N(X)$ will be either fully or partially explored. The updating of the tabu list is performed by making the reversal moves tabu and adjusting the tabu tenures of previous moves accordingly. The way this is carried out will be discussed later.

Algorithm 3.3: A Basic TS Heuristic

Step 1   Generate an initial solution, X, set the best current solution $X_{best} = X$, $F_{best} = F(X)$ and define the neighbourhood structure $N(X)$.

Step 2   Determine the best admissible solution $X' \in N'(X)$ where $N'(X) \subseteq N(X)$ and set $X = X'$ and update the tabu list. If $F(X') < F_{best}$ set $X_{best} = X'$ and $F_{best} = F(X')$.

Step 3   If an inner stopping criterion is met, go to Step 4, else go to step 2.

(continued)

Algorithm 3.3 (continued)

Step 4 (Diversification—optional) If an outer or global stopping criterion is not met yet, apply a diversification strategy based on X_{best} to obtain the new X and go to Step 2, else record the best solution X_{best}, its corresponding F_{best} and stop.

Basic Explanations of the Steps

Initial Solution This can be generated randomly, via a suitable heuristic or by an optimal method for a relaxed problem (here, the solution may not necessarily be feasible). The idea is to have a solution which can be easily suitable for the neighbouring search that is adopted.

Stopping Criteria (Steps 3 and 4) The inner stopping criteria can be related to the number of successive iterations without improvement, the maximum number of iterations in the inner loop, say Max($10n$, 1000) with n denoting the size of the problem. The stopping criteria of the outer loop can be defined by the maximum number of iterations, say Max ($100n$, 10, 000), the maximum time allowed, the maximum number of diversifications or the maximum number of successive diversifications without improvement. Note that if Step 4 is not activated the inner stopping criterion acts as the main stopping criterion instead.

Tabu List Size ($|T_s|$) This value aims to be small enough to approximately stop a solution from reoccurring. For instance, if the best admissible move is a non-improving move which was found by exploring the neighbourhood in its entirety, as one may expect, the reversal move in the next iteration will systematically select the previous better solution which will cause cycling. One way to stop such a cycle from occurring is by not allowing this reversal move to be accepted at least in the next iteration. But to be safe, we better have more than one as the configuration of the solution may not have changed enough. The value of $|T_s|$ can be an integer constant, a function based on the problem size or range-based like $|T_s| \in \mathbb{N} \cap [T_{\min}, T_{\max}]$. This can be randomly selected from the outset or chosen at each move. The choice of the $|T_s|$ can be critical to the final solution. Some attempts in modelling this critical factor will be discussed in the next subsequent subsection.

Tabu Restriction A recent accepted move is made tabu using the following representation. For instance, in routing problems where we have a number of routes and a number of customers involved, we can define this restriction as follows: Let $TAB(i, j)$ denote a matrix showing the tabu status of inserting customer i in route R_j. This can be defined by

(i) $TAB(i, j) = iter + |T_s|$ where *iter* is the current iteration count where the move is made tabu. In other words, customer i will be free to be reinserted back to route R_j only after $|T_s|$ iterations have passed.

(ii)
$$TAB(i, j) = \begin{cases} |T_s| & \text{once the move is accepted} \\ TAB(i, j) - 1 & \text{if } TAb(i, j) \geq 0 \text{ in subsequent iterations} \\ 0 & \text{otherwise} \end{cases}$$

The setting given in (i) is much quicker than (ii) as no updating is required.

Forbidding Strategy The move will remain tabu if
$$TAB(i, j) > \begin{cases} iter & \text{case (a)} \\ 0 & \text{case (b)} \end{cases}$$

Freeing Strategy A move becomes admissible

If $TAB(i, j) \leq iter$ (case (a)) or $TAB(i, j) \leq 0$ (case (b))
Else (the move is still tabu) $F(X') < F_{\text{best}}$ (case of minimisation). In other words, the new solution produces a better solution than what we have so far, and hence we obviously take it (i.e., aspiration overrides tabu restriction).

Note that the reason why this aspiration criterion is valid, in addition to the good quality solution, is that the setting of $|T_s|$ is experimentally set and not mathematically proven, and therefore it is used to guide the search only and not to blindly restricting it. This is an obvious but important observation which resides at the heart of heuristic search design in general. Other aspiration rules that are less used but in my view can be relatively more challenging will be mentioned later.

A Simple Illustrative Example Consider the example given by Glover (1990) which can easily be performed by hand. It is about finding the minimum spanning tree with some arc interdependency restrictions on a graph with five vertices and seven edges. An initial solution is first generated by Kruskal's greedy algorithm by ignoring the constraints. Although the solution happens to be infeasible, this solution is suitable for starting the TS mechanism and can also serve as a LB. Penalty values that are proportional to the amount of violation are added to the original objective function which corresponds to the length of the spanning tree. In this example, if edge i is selected to be introduced into the tree and another one, say j, is removed from the tree, $TAB(i)$ is set to $|T_s|$ being half of the tree size (i.e., 2). In other words, the ith edge is not allowed to be dropped from the tree during the next $|T_s|$ iterations. The value of $TAB(i)$ will be reduced by one at each iteration until it becomes non-tabu (i.e., $TAB(i) = 0$). Though the moves, the tabu restriction and the tabu size used are very simple, this example is easy to follow and demonstrates the progress of avoiding local minima very clearly.

TS Key Elements

Since TS has several key elements that could be worth examining, I would like to highlight those ones that I think to have potential. The reader can find the book on TS by Glover and Laguna (1997), and the work by Gendreau and Potvin (2010) to be interesting and informative.

Tabu Restrictions

How to define the right restriction? For instance, in routing, if the best move was to exchange customer i from route R_k with customer j from route R_s (not necessarily in the same positions), both customers will be made tabu. Ways of restricting these two customers may include the following:

(a) A very tight restriction (both customers cannot be allowed to go back to their original routes respectively for a certain number of iterations).

(b) A tight restriction: Customer i or j is not allowed to go back to its original route but not both. The question here is which is which.

(c) A less tight restriction: Both customers are not restricted to go back to their original routes but cannot be allowed to be inserted between their original predecessors and successors. This explains that the route elements may have been changed, and therefore the most important factor is their original respective positions which need to be made tabu.

It should be noted that even if a complete configuration is allowed to be regenerated, the probability of cycling may not be exactly equal to one due to the tabu status of the other attributes at that given iteration. Allowing such a restriction to be a little more relaxed may contribute to the understanding of the way tabu restrictions are defined.

Tabu List Size $|T_s|$

Mall values of $|T_s|$ may turn the search to become too flexible and therefore may increase the risk of cycling, whereas large values on the other hand may constrain the search to work on narrow regions where many neighbours are tabu active. The ideal value may not exist, however the following definition could help. We can define $|T_s|$ either in a static or in a dynamic way so to reflect the above considerations.

Constant $|T_s|$

This can take a fixed value, or defined from the outset, say $\frac{n}{\theta}$ where n is the size of the problem (number of cities for the TSP, number of vertices in a tree, number of customers in a location problem) and $\theta \in \{2, 3, ..., 6\}$ is a control parameter. For instance, Osman and Salhi (1996) produced competitive results using $\theta \in \{4, 5, 6\}$ for the vehicle routing problem with heterogeneous vehicle fleet.

Dynamic Changes in $|T_s|$

(i) *Periodically changing* $|T_s| \in [T_{\mathrm{Min}}, ..., T_{\mathrm{Max}}]$
This can be generated randomly in this range either at each iteration, or once in a while, say K_{max} times (i.e., once generated it is kept the same for a certain number of iterations). The choice of T_{min}, T_{max} and K_{max} can be critical to the success of the search. This is known as robust tabu as it provides flexibility. For instance, for the case of the quadratic assignment problem Skorin-Kapov (1990) produced   successful   results   using $T_{\mathrm{min}} = 0.9n, T_{\mathrm{max}} = 1.1n$ and $K_{\mathrm{max}} = 2T_{\mathrm{max}}$ with n being the number of facilities. Some attempts were also made for the p-median problem using the following values $K_{\mathrm{max}} = 3$, $T_{\mathrm{min}} = \frac{p}{3}$ and $T_{\mathrm{max}} = \frac{2p}{3}$ with p being the number of facilities that need to be opened (Salhi 2002).
Another scheme is to have an alternating $|T_s|$ based rule that switches between the two extreme values of T_{min} and T_{max}. This implementation was found to be successful when adopted by Drezner and Salhi (2000) for the one way network design problem.

(ii) *Continuously changing* $|T_s|$

$$|T_s| = \begin{cases} H(\Delta) & \text{if } \Delta \geq 0 \text{ (positive improvement)} \\ T_{\mathrm{Min}} & \text{otherwise} \end{cases}$$

where $H(\Delta)$ is a smooth strictly increasing function, T_{min} the minimum threshold value of $|T_s|$ and $\Delta = F(X') - F(X)$ with $X' \in N(X)$.

For instance, for the p-median problem, the following functional setting of $|T_s|$ was successfully adopted by Salhi (2002).

$$|T_s| = \begin{cases} \text{Min}\{T_{\text{Max}}, T_{\text{Min}} + e + \text{Log}(1 + \Delta)\} & \text{if } \Delta \geq 0 \\ T_{\text{Min}} + e^{(1+\Delta)} & \text{otherwise} \end{cases}$$

where $T_{\min} = [p/3]$ and $T_{\max} = p - 1$ with p being the number of facilities to open.

The construction of such functions is seldom attempted in the literature and any development along this line of research would, in my view, be not only challenging but interesting as this incorporates both learning and problem characteristics in an integrated way. This can be dependent on the change in the cost function for that selected move.

(iii) *Reactively changing $|T_s|$*

The size of the tabu list can also be updated dynamically by increasing or decreasing its value as the search progresses. Usually, $|T_s|$ is increased ($|T_s| = (1 + \beta)|T_s|$) with $\beta = 0.1$ or decreased by $\beta = -0.1$ depending on the number of repetitions, their risk of collision and so on. Collision is usually identified through hashing functions or other forms of identification. This is originally proposed by Battiti and Tecchiolli (1994) who named this variant as reactive TS (RTS). This approach is adopted by few researchers including Wassan (2006) who successfully addressed a class of routing problems, namely, the vehicle routing with backhauls. RTS is shown to be among the best TS procedures due to focussing on getting a suitable balance between intensification and diversification. In addition to the dynamic update of $|T_s|$, RTS differs also from the standard TS due to its reliance of a large number of random moves which in many cases will provide an opportunity for the search to get out of local optima. The design of the hashing function can be challenging which I will revisit in the implementation chapter.

Soft Aspiration Level

The level for which aspiration overrides tabu status is crucial in the search as this defines the degree of flexibility of the method. One common sense-based scheme is to relax the tabu restriction if a solution happens to produce a better result than the currently best known solution. This is obviously correct as the search would look really blind if it was that rigid.

A more advance criterion is what will happen if all the top solutions are both tabu and non-improving solutions? Is it appropriate to look for the first non-tabu solution down the list or to choose one from those tabu top solutions? The elements which will constitute the decision need to include the tabu status of the attribute for that solution, the objective function value (or the change in the objective function) and other factors such as frequency of occurrence during a certain number of iterations and so on. One possible attempt is to use the following:

Consider at a given iteration, M solutions are found and all of them happen to be non-improving solutions. For simplicity, let us put these solutions in a list of ascending order of their change in the objective function from the best (i.e., $\Delta_i = {}^i F(X') - F(X)$) and assume that the first top q solutions are all tabu.

$$\begin{pmatrix} S_1 & \Delta_1 \\ S_2 & \Delta_2 \\ - & - \\ S_q & \Delta_q \\ - & - \\ S_M & \Delta_M \end{pmatrix}$$

In brief, let

$S_1, \ldots, S_q, \ldots, S_M$ be all non-improving solutions and $S_1, \ldots, S_q$ tabu. $\Delta_1, \ldots, \Delta_M$ the change in the objective function value with respect to the current best value with $\Delta_1 \leq \ldots \leq \Delta_q \leq \ldots \leq \Delta_M$.

If $\Delta_1 < 0$ S_1 is an improving solution and S_1 would be chosen according to the usual aspiration criterion by overriding the tabu status. As this is not the case given that $\Delta_1 \geq 0$, the choice is to either

(i) take the first move in the list that is not tabu, say S_{q+1} or
(ii) introduce a selection rule for choosing $S_y \in \{S_1, \ldots, S_q\}$.

In (i) the usual way of using the tabu restriction rule is strictly applied, whereas in (ii) more flexibility is introduced. The latter which is more challenging can be approached as follows through a softer criterion:

Softer Aspiration Level

This criterion is based on the concept of criticality in the tabu status (e.g., an attribute which has a tabu tenure of 10 is more 'tabu' than the one associated with a value of 1 or 2.) and the closeness of the solution quality (a solution may be closer to the best current solution but it still non-improving).

Definition A reversal move j is said to be α_j tabu if there are α_j iterations left for move j to become non-tabu.

Based on the above reasoning, the following selection rule can be used:

$$\delta_j = \frac{(\Delta_j - \Delta_q)c_1}{\alpha_j^{c_2}} \text{ and } \alpha_j \geq 1 \text{ for } j = 1, \ldots, q$$

where
$\alpha_j = TAB(j) - iter$ is the tabu tenure of attribute j (the number of iterations remaining for attribute j to become non-tabu, i.e., attribute j is α_j tabu), and c_1 and $c_2 \in [0, 1]$ are parameters describing the importance of penalising a move with a large change in objective function and small tabu tenure, respectively.

The move to be selected, say j^*, is the one that yields the largest δ_j; in other words, $j^* = \text{ArgMax}(\delta_j j = 1, \ldots, q)$.

For example, the values of $c_1 = 1$ and $c_2 = 0.5$ have shown empirically to produce good results for the p-median problem (see Salhi 2002). The above rule could be modified to consider those moves with $\alpha_j \leq \alpha_{max}$ only.

Diversification and Intensification

Diversification The aim is to provide moves that make the next state (neighbourhood) differs significantly from the current neighbourhood. Diversification moves use some form of perturbation modules (jumps) to explore new regions. It uses long-term memory to guide the search by exploring other search spaces that have not been visited yet. This is achieved either by moving out of a region which appears to be less promising according to the results provided from the short-term memory, or by identifying other areas that ought to be considered. In other words, the idea is to try to cover as much as possible the search space. For instance, in TS, this is performed mainly through recording the changes in the tabu status of the attributes by using frequency of occurrence of an attribute in a selected solution. The use of diversification is nowadays embedded into most metaheuristics either as a post optimisation step or as a perturbation step as in LNS or in the iterated local search.

Intensification This stimulates moves that go to a nearby state (neighbourhood) that looks myopically good. This uses short-term memory as it observes the attributes of all performed moves. This is usually achieved by a local search or a series of local searches where the focus of the search is on those so called important or promising regions. In brief, intensification is aimed at detecting good solutions and performing a deeper exploration.

The following strategies are usually used either at intensification or diversification levels or even both:

  (i) Restarting the search from high quality solution found at earlier iterations
 (ii) Aggregating attributes that have been consistently sequenced in the same order during the course of the search
(iii) Locking some attributes in the best (or current) solution into their best (or current) positions and performing the search on the free attributes

(iv) Using frequency information to construct a starting solution with attributes occupying the preferred positions

The question which is important is how to decide when it is better to carry out an intensification or it is time to perturb the solution and activate diversification? In other words, how do we measure the relative solution quality to enable more or less intensification around the current neighbourhood solution? How to distort the structure of the current solution (is it randomly or via intelligent search)? Do we want to revisit good regions as well as exploring those unvisited ones? What is the right balance between intensification and diversification? None of these questions are straightforward as these are related to the problem characteristics, the power of the local search used and the type of perturbation the overall search is adopting.

Strategic Oscillation (SO)

TS is a search that needs to be flexible and capable to cross boundaries between feasible regions. This concept is useful particularly when the solution space is disconnected or none convex as it is impossible to cross the infeasible region using only feasible solutions. In addition to guiding the search from not diverging, a penalty function that is dynamically adjusted is usually incorporated as part of the objective function to handle the constraints, see Gendreau and Potvin (2010) for more details. As an example, an initial implementation was presented by Kelly et al. (1993) who successfully integrated TS and SO when solving a class of routing problems. Another useful application of TS and SO was conducted successfully by Hanafi and Freville (1998) when solving a class of 0–1 multi-dimensional knapsack problems. Their TS was efficiently designed to shift between feasible and infeasible regions in a guided manner. Note that this cross-boundaries shortcoming could become less severe if larger jumps are incorporated into the search as in LNS, for example. As this constraint handling aspect can be part of any powerful heuristic search that incorporates infeasibility within its search strategy, I shall revisit this aspect in the implementation chapter.

Some Thoughts

TS is a powerful metaheuristic that relies on several key elements some have shown to be promising. The tabu size is still one of the factors not explored fully with the exception of some attempts in RTS. A functional-based $|TS|$ is rarely explored but can be worth revisiting. The way the aspiration criterion is adopted is very simple and adopting a softer aspiration criterion, as attempted by Salhi (2002), though challenging could be worth exploring. Some work on the convergence of SA exist but there is relatively little attempt in addressing a similar theoretical issue in TS with the exception of the interesting study by Glover and Hanafi (2002) who attempted to demonstrate the finite convergence of TS under certain assumptions. Some theoretical work along this line could be interesting and mathematically challenging.

3.4   Summary

The three heuristics, namely, SA, TA and TS are presented. These are powerful metaheuristics that accept non-improving solutions through controlling their search. A probabilistic acceptance is adopted in SA, deterministic thresholding in TA and tabu moves in TS. The success of these modern heuristics relies heavily on the neighbourhood chosen, the selection rules adopted and the way their key elements are defined and implemented. Some of these main components are discussed and their effect on the search highlighted. The next chapter covers the case of using more than one solution at a time.

References

Aarts, E. H. L., & van Laarhoven, P. J. M. (1985). Statistical cooling: A general approach to combinatorial optimization problems. *Philips Journal of Research, 40*, 193–226.

Battiti, R., & Tecchiolli, G. (1994). The reactive tabu search. *ORSA Journal on Computing, 6*, 126–140.

Conolly, D. T. (1990). An improved simulated annealing technique for the QAP. *European Journal of Operational Research, 46*, 93–100.

Dowsland, K. A. (1993). Some experiments with simulated annealing techniques for packing problems. *European Journal of Operational Research, 68*, 389–399.

Dowsland, K. A., & Thompson, J. M. (1998). A robust simulated annealing based examination timetabling system. *Computers and Operations Research, 25*, 637–648.

Dowsland, K. A., & Thompson, J. M. (2012). Simulated annealing. In G. Rozenberg, T. Back, & J. N. Kok (Eds.), *Handbook of natural computing* (pp. 1624–1655). Berlin: Springer.

Drezner, Z., & Salhi, S. (2000). Using tabu search for designing one and two ways road networks. *Control and Cybernetics Journal, 29*, 725–740.

Drezner, Z., & Salhi, S. (2002). Using hybrid metaheuristics for the one-way and two-way network design problem. *Naval Research Logistics (NRL), 49*, 449–463.

Drezner, Z., Marcoulides, G. A., & Salhi, S. (1999). Tabu search model selection in multiple regression analysis. *Communications in Statistics Simulation and Computation, 28*, 349–367.

Dueck, G. (1993). New optimization heuristics: The great deluge algorithm and the record-to-record travel. *Journal of Computational Physics, 104*, 86–92.

Dueck, G., & Scheuer, T. (1990). Threshold accepting: A general purpose optimization algorithm superior to simulated annealing. *Journal of Computational Physics, 90*, 161–175.

Eglese, R. (1990). Simulated annealing: A tool for operational research. *European Journal of Operational Research, 46*, 271–281.

Gendreau, M., & Potvin, J. Y. (2010). Tabu search. In M. Gendreau & J. Y. Potvin (Eds.), *Handbook of metaheuristics* (pp. 41–59). London: Springer.

Glover, F. (1986). Future paths for integer programming and links to artificial intelligence. *Computers and Operations Research, 13*, 533–549.

Glover, F. (1990). Tabu search: A tutorial. *Interfaces, 20*, 74–94.

Glover, F., & Hanafi, S. (2002). Tabu search and finite convergence. *Discrete Applied Mathematics, 119*, 3–36.

Glover, F., & Laguna, M. (1997). *Tabu search*. Boston: Kluwer.

Hanafi, S., & Freville, A. (1998). An efficient tabu search approach for the 0–1 multidimensional knapsack problem. *European Journal of Operational Research, 106*, 659–675.

Hansen, P. (1986). The steepest ascent, mildest descent heuristic for combinatorial programming. *Paper presented at the congress on Numerical Methods in Combinatorial Optimization*, Capri.

Hu, T. C., Kahng, A. B., & Tsao, C. W. A. (1995). Old bachelor acceptance: A new class of non-monotone threshold accepting methods. *ORSA Journal on Computing, 7,* 417–425.

Johnson, D. S., Aragon, C. R., McGeoch, L. A., & Schevon, C. (1989). Optimization by simulated annealing: An experimental evaluation. Part I, graph partitioning. *Operations Research, 37,* 865–892.

Kelly, J. P., Golden, B., & Assad, A. A. (1993). Large-scale controlled rounding using tabu search with strategic oscillation. *Annals of Operations Research, 41,* 69–84.

Kirkpatrick, S., Gelat, C. D., & Vecchi, M. P. (1983). Optimization by simulated annealing. *Science, 220,* 671–680.

van Laarhoven, P. J. M., & Aarts, E. H. L. (1987). *Simulated annealing: Theory and applications.* Rotterdam: Reidel.

Lee, D. S., Vassiliadis, V. S., & Park, J. M. (2004). A novel threshold accepting meta- heuristic for the job-shop scheduling problem. *Computers and Operations Research, 31*(13), 2199–2213.

Li, F., Golden, B., & Wasil, E. (2007). A record-to-record travel algorithm for solving the heterogeneous fleet vehicle routing problem. *Computers and Operations Research, 34,* 2734–2742.

Lundy, M., & Mees, A. (1986). Convergence of an annealing algorithm. *Mathematical Programming, 34,* 111–124.

Metropolis, N., Rosenbluth, A., Rosenbluth, M., Teller, A., & Teller, E. (1953). Equations of state calculations by fast computing machines. *The Journal of Chemical Physics, 21,* 1087–1092.

Osman, I. H., & Christofides, N. (1994). Capacitated clustering problems by hybrid simulated annealing and tabu search. *International Transactions in Operational Research, 1,* 317–336.

Osman, I. H., & Laporte, G. (1996). Metaheuristics: A bibliography. *Annals of Operations Research, 63,* 513–623.

Osman, I. H., & Salhi, S. (1996). Local search strategies for the vehicle fleet mix problem. In V. J. Rayward-Smith, I. H. Osman, C. R. Reeves, & G. D. Smith (Eds.), *Modern heuristic search techniques* (pp. 131–154). New York: Wiley.

Salhi, S. (2002). Defining tabu list size and aspiration criterion within tabu search methods. *Computers and Operations Research, 29,* 67–86.

Skorin-Kapov, J. (1990). Tabu search applied to the quadratic assignment problem. *ORSA Journal on Computing, 2,* 33–45.

Tarantilis, C. D., Kiranoudis, C., & Vassiliadis, V. (2003). A list based threshold accepting metaheuristic for the heterogeneous fixed vehicle routing problem. *The Journal of the Operational Research Society, 54,* 65–71.

Wassan, N. A. (2006). A reactive tabu search for vehicle routing. *The Journal of the Operational Research Society, 57,* 111–116.

4

Population-Based Heuristics

4.1 Genetic Algorithms

Genetic algorithms (GAs) were initiated to mimic some of the processes observed in natural and biological evolution (Darwinian theory: *survival of the fittest*). These processes are then organised in a structured manner to make up a step-by-step algorithm. This approach was initially developed by John Holland and his associates at the University of Michigan in the 1970s. It was originally used for game theory and pattern recognition problems. It was only in the 1975 when the method was formally introduced for the context of optimisation in general and combinatorial optimisation in particular; see Holland (1975) and Goldberg (1989).

In brief, GA is an adaptive heuristic search method based on population genetics. GA consists of a population of chromosomes (set of solutions) that evolve over a number of generations (iterations) and are subject to genetic operators (transformations) at each generation. Each chromosome has a fitness value associated with it (the objective function or one of its related functions), with the view that the best fit chromosomes have more chance to survive from one generation to the next (survival of the fittest). The main steps of a basic GA are given in Algorithm 4.1.

© The Author(s) 2017
S. Salhi, *Heuristic Search*, DOI 10.1007/978-3-319-49355-8_4

Algorithm 4.1: A Basic Genetic Algorithm

Step 1   Initialise a population of chromosomes.
Step 2   Evaluate each chromosome of the population.
Step 3   Create new chromosomes by mating current chromosomes using suitable operators.
Step 4   Evaluate the new chromosomes
Step 5   Delete some chromosomes from the augmented population (or from the old one only) to maintain the size of the population.
Step 6   If certain stopping criteria are met, stop; otherwise go to Step 3.

In brief, GA which is a population-based approach differs significantly from other already discussed heuristics which rely on a one point solution as it considers a set of solutions simultaneously instead. The main idea is that evolution takes place on chromosomes and there are chromosomal encoding and decoding processes to generate the new chromosomes. These processes are described next.

Chromosome Representation

The population members are string entities of artificial chromosomes. For instance, in facility location such as the p-median problem, the aim is to locate p facilities (say 5) out of n potential sites (say 50, each located at a customer site for simplicity) so that the total cost of serving these customers from those p open facilities is minimised. As a possible configuration, consider the chosen facilities to be at sites 2, 4, 25, 47 and 49. The chromosome can be represented either

(a) in a binary encoding where a string of size 50 is shown like in Fig. 4.1. Here, each bit is binary denoting whether or not a site at that position (i.e., from 1 to 50) is selected or not with value 1 being selected and 0 otherwise or

(b) as an integer encoding where a chromosome of size 5 is defined with each bit representing the location number of the open site as shown in Fig. 4.2.

0	1	0	1	0	0... 0 1 0 ... 0	1	0	1	0

Fig. 4.1 A binary encoding of the five facilities example

2	4	25	47	49

Fig. 4.2 An integer encoding of the five facilities example

Both representations are valid but for this type of application, the second is simpler and more appropriate as it easily caters, for instance, with a large number of customers (acting as potential sites) and a reasonable number of facilities (i.e., $n = 1000$ and $p = 20$). Besides, it is relatively easier to handle operators such as exchanging facilities sites or adding new/dropping old facilities. However, it is worth noting that the latter may generate infeasible offsprings (i.e., the same facility is chosen more than once) such infeasibility is easily repaired as will be shown later in this section. This flexibility of selecting the right representation demonstrates the need to be cautious when choosing the chromosome representation as this can either facilitate or hinder the chromosomes transformation that will be generated via the GA operators which I shall discuss next.

Integer/Real Numbers Representation

Note that any integer number can be written in a decimal system (e.g., in base 10). For instance, the variable $x = 39$ can be represented as a string of length of 2 as $x = (3, 9)$ in base 10 as its value corresponds to $3.10^1 + 9.10^0$. Similarly, it is also possible to code it in a binary form (e.g., in base 2) using a string of size 6 bits as follows $x = (100111)$ as $x = 39 = 1.2^5 + 0.2^4 + 0.2^3 + 1.2^2 + 1.2^1 + 1.2^0$.

Real valued or integer numbers are usually represented in binary string. Real values can be transformed from integer ones by using the following: Take the example of $F(X)$ where $X \in [a, b]$.

– Generate a bit string of length k, say 22. For instance, this gives X' = $(01011\ldots.0110)$ with $X' \in [0, 2^{22} - 1]$.

– Translate X' into $X \in [a, b]$ by defining $X = a + X' . \frac{(b-a)}{(2^{22}-1)}$.

– Compute $F(X)$ and repeat the process to construct the other parents.

GA Operators

The commonly used GA operators include mutation, inversion, crossover and reproduction. The first three are performed to transform the current chromosomes into new ones, whereas the last one focuses on the selection and removal of parents.

Bit Mutation

For each bit in each string, a random number is generated, say α. If $\alpha \leq \beta$ (a fixed acceptance probability value, say $\beta = 0.005$, usually this value tends to be close to 0), then the value of that bit can be considered for possible mutation. For instance, in our location example, as represented in Fig. 4.2, the bit may or may not change its original value depending whether or not the mutation is passed. If it is, a random site among all the non-chosen potential sites is chosen for that bit. The selection of parents for mutation is a question which is critical to the success of GA as this may consume unnecessary additional computational effort. One way would be to either randomly choose the parents based on certain rules and then all, or some, of their bits are randomly selected.

Inversion

This operator works on a single chromosome. It generates two positions in the string randomly or via a process, and then all the elements between those two points, separated by *, will have their bit values inverted (from 0 to 1 and vice versa) as shown in the following illustrative example.

Parent 1 : (100 * 1001* 1) → Child 1 : (10001101)

If the other representation is adopted, each bit in between the two crossover points is randomly exchanged with another number not in the current chromosomes. This operator in this situation reduces to using a successful mutation for each bit.

It is worth noting that such an operator tends to perturb the current solution drastically, and therefore it could be used once in a while for diversification purposes rather than regularly.

Crossover Operators

In nature, crossover occurs when two parents exchange parts of their corresponding chromosomes. There can be based on one point or a multiple points crossover.

One Point Crossover

For illustration, consider first the simple one point crossover which aims to swap all the bits of two parent chromosomes (chosen randomly or intelligently) after a given selected point to form two children.

For instance, consider the following two chromosomes represented by strings of 5 bits and assume that the crossing point is after the second position in the string. This leads to two children which differ from their parents as shown below.

$$\begin{cases} \text{Parent1}: & (01\,|001) \rightarrow \text{Child1}: & (01100) \\ \text{Parent2}: & (10\,|\,101) \rightarrow \text{Child2}: & (10001) \end{cases}$$

Consider the location example given in Fig. 4.2 where Parent 1 and Parent 2 will be mated using the crossover point after the second bit. This leads to two children as shown below.

$$\begin{cases} \text{Parent1}: & (2-4\,|25-47-49) \rightarrow \text{Child1}: & (2-4-30-40-45) \\ \text{Parent2}: & (15-20\,|\,30-40-45) \rightarrow \text{Child2}: & (15-20-25-47-49) \end{cases}$$

However, in some cases, the mating may generate infeasible children as shown in the following example where both children have duplicate facilities (i.e., facility # 15 in Child 1 and facility # 25 in Child 2).

$$\begin{cases} \text{Parent1}: & (10-15\,|25-40-45) \rightarrow \text{Child1}: & (10-15-15-30-43) \\ \text{Parent2}: & (20-25\,|\,15-30-43) \rightarrow \text{Child2}: & (20-25-25-40-45) \end{cases}$$

This can be easily repaired by either (a) restricting the mating to those pairs of parents that do not produce such duplications (in other words, reject such children and repeat the mating process), or (b) introduce a repair mechanism to alleviate such infeasibility. The latter is used more frequently as it is deterministic, whereas the first may require too many computations and it may even lead to impossible mating.

A repair mechanism is to construct two subsets of the free facilities for each parent by excluding those that are already in the first part of the chromosome before the crossover point. First, let us record those facilities that are in common to both parents say $E_c = P_1 \cap P_2$ with P_1 and P_2 denoting the facility configuration for Parent 1 and Parent 2, respectively. For our location example, we have $E_c = \{15, 25\}$ and let the remaining facilities belong to the subset $\bar{E} = P_1 \cup P_2 \backslash E_c$. Also consider P_1^1 and P_2^1 be the first parts of the chromosome before the crossover point of P_1 and P_2 respectively, and let $\bar{E}_1 = \bar{E} \backslash \bar{E} \cap P_1^1$ and $\bar{E}_2 = \bar{E} \backslash \bar{E} \cap P_2^1$. The two children are then formed as Child $1 = E_c \cup P_R^1$ and Child $2 = E_c \cup P_R^2$ with the facilities in P_R^1 and P_R^2 being randomly generated from $\bar{E}_1$ and $\bar{E}_2$,

respectively. In our example, the two children are formed as follows: Child
$1 = (10 - 15 - a - b - c)$ and Child 2 $= (20 - 25 - a' - b' - c')$
with (a, b, c) and (a', b', c') being randomly generated from
$\bar{E}_1 = \{25, 40, 45, 20, 30, 43\}$ and $\bar{E}_2 = \{15, 30, 43, 10, 15, 45\}$,
respectively.

A simpler repair mechanism which keeps the crossover characteristics as
much as possible is to consider the infeasible children and swap the
duplicate facilities with the following as follows:

Child 1 : $(10 - 15 - a - 30 - 43)$ and Child 2 : $(20 - 25 - b$
$-40 - 45)$ with a and b initially set to duplicated facilities #15 and
25, respectively, which need to be changed.

Let $\widehat{E}_1 = \bar{E}_1 \backslash \bar{E}_1 \cap P_2^2$ and $\widehat{E}_2 = \bar{E}_2 \backslash \bar{E}_2 \cap P_1^2$ with P_2^2 and P_2^2 denoting
the second part of the chromosome in Parent 2 and Parent 1, respectively.
In our example, we have $\widehat{E}_1 = \{40, 25, 45, 20\}$ and $\widehat{E}_2 = \{30, 3, 15, 10\}$.
We therefore randomly choose a from $\widehat{E}_1$ and b from $\widehat{E}_2$ to make up two
new feasible chromosomes.

It is worth mentioning that the above mechanisms have some similarities with the repair operator originally presented by Alp et al. (2003) for
the $p-$ median problem where the two children are formed as Child 1
$= E_c \cup P_R^1$ and Child 2 $= E_c \cup P_R^2$ with P_R^1 and P_R^2 referring to the subset
of facilities that are randomly chosen from $\bar{E}$. These two subsets can be
constructed either separately and independently from each other (leading
to a few facilities being the same for both children), or dependently by
deleting any randomly selected facility from $\bar{E}$ once it is chosen. In this
example, the following schema for the two children would have been
obtained:

Child 1 $= (15 - 25 - a - b - c)$ and Child 2 $= (15 - 25 - a'$
$-b' - c')$ with (a, b, c, a', b', c') being randomly chosen from
$\bar{E} = \{10, 40, 45, 20, 30, 43\}$.

Multi Points Crossover

In some situations it is too restrictive to have only one-point crossover as all the elements after the crossover point are swapped with those of the other parent. It is rather useful to swap a part of those elements of the parents only. To do that at least two-point crossover needs to be used. The number of crossover points may vary from problem to problem, though usually using one- or two-point crossovers are the most common.

Consider the case of two points crossover at position 2 and 4 for the following two parents of six bits where the bits of the parents in between the two points are swapped.

$$\begin{cases} \text{Parent } 1 : \left(01|00|10\right) \rightarrow \text{Child} 1 : \left(011110\right) \\ \text{Parent } 2 : \left(10\,|\,11|01\right) \rightarrow \text{Child} 2 : \left(100001\right) \end{cases}$$

PMX Operator

It is worth noting that the above implementation may fail. For instance, in some combinatorial problems, the solutions are represented by sequences as in the TSP where an obvious coding representation is to put cities number in the bit of the strings. The ordinary crossover is not appropriate as cycles are likely to occur. Consider the example of 7 cities where two random tours are generated and the crossover point is at position 3. This operation may lead to infeasible solutions as there are city duplicates (as shown in the previous example of the $p-$ median problem) and some cities are not even visited.

$$\begin{cases} \text{Parent} 1 : \left\{7-6-5-|4-3-2-1\right\} \\ \quad \rightarrow \text{Child} 1 : \left\{7-6-5-5-7-6-1\right\} \\ \text{Parent} 2 : \left\{1-3-4-|5-7-6-2\right\} \\ \quad \rightarrow \text{Child} 2 : \left\{1-3-4-4-3-2-2\right\} \end{cases}$$

To overcome this drawback, Goldberg and Lingle (1975) developed a partially mapped crossover (PMX for short) which avoids cycles. This is performed as follows:

PMX operator uses two points crossover and a mapping neighbouring scheme.

For ease of illustration consider the same example of the seven cities. The following possible mapping is then used.

$$\left. \begin{array}{l} 4 \leftrightarrow 5 \\ 3 \leftrightarrow 7 \\ 2 \leftrightarrow 6 \end{array} \right\} \Rightarrow \left\{ \begin{array}{l} \text{Child 1} : \{3 - 2 - 4 - 5 - 7 - 6 - 1\} \\ \text{Child 2} : \{1 - 7 - 5 - 4 - 3 - 2 - 6\} \end{array} \right.$$

Such a mapping is suitable as long as there is a one-to-one relationship in mapping. For instance, consider the following two parents.

$$\left\{ \begin{array}{l} \text{Parent 1} : \{7 - 6 - 5 - 4 - 3 - 2 - 1\} \\ \quad \rightarrow \text{Child 1} : \{3 - 2 - 5 - 5 - 7 - 6 - 1\} \\ \text{Parent 2} : \{1 - 3 - 4 - 5 - 7 - 6 - 2\} \\ \quad \rightarrow \text{Child 2} : \{1 - 4 - or\, 7 - 4 - 3 - 2 - 6\} \end{array} \right.$$

The mapping used is: $4 \leftrightarrow 3; 3 \leftrightarrow 7; 2 \leftrightarrow 6$.

This means that 3 is mapped to both 4 and 7. Note that when applying this mapping, Child 1 still fails to satisfy the cycle property whereas Child 2 is not uniquely defined. There are, however, ways to overcome such a shortcoming which can be found in Goldberg and Lingle (1975).

Reproduction/Selection Mechanisms

This is a process where certain strings (solutions) are copied fully into the next generation, usually the fittest (the best ones). Note that reproduction aims to keep a number of the top best current chromosomes from generation to another, the most commonly approach is to keep the best which is usually referred to as 'elitism'. The size of the new population is usually maintained using the $(m + n)$ strategy where m is the size of the parent population and n refers to the number of children, and among the $m + n$ chromosomes, the best chromosome is first chosen and the remaining $m - 1$ are then selected pseudo-randomly based on certain selection rules. This concept of elitist model is well documented by De Jong (1975).

The way to select which parent to drop or to mate with another for further experimentation is a critical issue that needs closer look. Is it simpler to discard a fraction of the below average fitness, and maintain the top fittest in the population from one generation to the next? Does this strategy need to be implemented from the beginning of the search, or is that fraction dependent on the generation number and on the average fitness? For the purpose of retaining diversification in early stages, the search is less limited and becomes more and more restrictive as the solutions become closer and closer to the local optimum.

Tournament Selection

A commonly used selection scheme is the tournament selection where a subset of the population is chosen based on their fitness values (i.e., $f(.)$). There are several ways to select a chromosome k from a population P for reproduction. The obvious one is the fitness proportional selection based on the probability of selecting chromosome k with $\text{Prob}(k) = f(k)/\sum_{r \in P} f(r)$. This is the roulette wheel selection which is commonly used. This simple rule seems to suffer from those dominant chromosomes especially at the earlier generations as this process tends to reach early convergence making the search relatively slower at the end. Other selection rules are based on linear ranking, power and exponential ranking, among others.

Matching Issues

How to match the parents can also play an important part. It is useful to distinguish between the overall fitness of a chromosome and its impact if combined with another chromosome. The concept of building block is crucial if it can be explicitly identified since a small building block from a less fit chromosome can be exceptionally good if inserted into the already fit chromosome. The opposite can also happen if a good chromosome loses its attractiveness through one of its building blocks and gains a poor one from another parent.

Flexible Reproduction

One interesting way would be to sort the chromosomes according to their fitness, then construct in addition to the top chromosome, three groups consisting of good, not so good and mediocre or bad chromosomes with respective sizes n_1, n_2 and n_3, respectively. The weights of the three respective groups at generation t say $\alpha_1(t) > \alpha_2(t) > \alpha_3(t) > 0$ with $\sum_{j=1}^{3} \alpha_j(t) = 1$ can be defined. We can also assume that the first weight $\alpha_1(t)$ is a non-decreasing function of t with $\lim_{t \to \infty} \alpha(t) \to 1$ while the other weights converge towards zero. Using the roulette wheel, the group will be selected based on these weights and then a chromosome within that group is chosen randomly for crossover or mutation and so on. For the case of chromosomes removal, the reverse can be adopted.

Effect of Migration

It is also important to provide diversity, and hence an opportunity for improvement through some form of migration. In GA, a small number of completely new chromosomes which are generated either randomly or constructed can be injected into the population to provide diversity once in a while. The number of injected chromosomes and when the injection takes place are issues that deserve careful investigation. Usually the injection starts being active once the GA shows some form of stagnation and then from that point onwards injection is activated periodically or adaptively depending on the behaviour of the overall results.

Figure 4.3 illustrates the new combined generation scheme which includes both the flexible reproduction and the effect of immigration. A similar approach was successfully explored by Salhi and Gamal (2003) and Salhi and Petch (2007) for a class of location and routing problems, respectively.

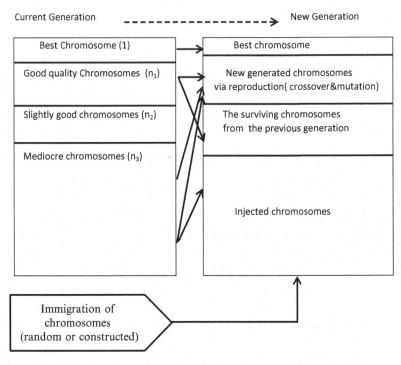

Fig. 4.3   Injection of chromosomes and new-generation composition

Fitness Function Representation

The representation of the fitness function is not as easy as it sounds. I present the following fitness functions which can be also used as a basis for other modifications if necessary.

(i) The obvious fitness function is the use of the raw objective function. This representation can be misleading if some chromosomes outperform by far others at early stages of the search. This causes premature convergence which leads to local optimality as many parents will be the same or very similar. In such circumstances, inversion could be a key operator acting as a powerful diversification strategy to overcome such a limitation as it has the flexibility in creating diverse chromosomes.

(ii) One way to overcome the drawback of (i) is to introduce a scaling factor to give relatively less weight to the fittest chromosomes making the selection process less bias towards considering only those fitter chromosomes. This scaling is important at the beginning of the search or when the search seems to get stuck at local optima. In other words, it limits competition at early stages of the simulation but stimulates the competition at later stages.

The following scaling function is usually applied:

$$f = \alpha F + \beta$$

where f is the fitness, F the original objective function and α and β parameters whose values are determined using the following conditions: $\bar{f} = \bar{F}$ and $f_{\text{Max}} = v\bar{F}$ with v used as a control parameter to ensure that the fittest chromosome of the population will be chosen on average v.

(iii) Instead to base the decision on their objective function values as in (i) or (ii), potential parents are selected using a ranking procedure based on the probability distribution $P(k) = \frac{2k}{m(m+1)}$ where k denotes the kth chromosome ranked in ascending order of $F(.)$ and m is the number of chromosomes in the population. This rule gives more emphasis to the ranking order of the chromosomes with regard to the objective function value rather than the value of the objective function itself.

(iv) Incorporate a penalty function to aid the GA in accepting infeasible solutions so to widen its search space. A penalty function based on the amount of infeasibility is usually used. This aspect will be revisited in the implementation chapter when dealing with constraint handling.

Some Thoughts

Successful implementations of GA rely on several aspects including (i) how to better represent a chromosome, (ii) how to design appropriate operators with or without repair mechanisms, (iii) how to select intelligently the parents to mate and (iv) how to develop schemes to avoid early convergence. The latter can be achieved by not ignoring completely those parents that are considered of a poor quality (mediocre ones) and also by allowing injection of new chromosomes into the population either randomly, dynamically or adaptively. GA is powerful at exploring the wider space by identifying promising regions, but it lacks the fine mechanisms to pinpoint the exact local minima (maxima) that may be required. This issue of integration will be discussed next in the hybridisation chapter.

4.2   Scatter Search

Scatter search (SS), proposed by Glover et al. (2003), is also an evolutionary approach that constructs new solutions by combining bits (parts) of the already found solutions. In brief, this population-based method is more flexible than the GA when it comes to constructing parents from children. The idea is to generate a large number of diverse trial solutions which are then improved. The set of these solutions needs to be of high quality while being diverse. There are many ways of implementing SS but a commonly used approach is to rely on five schemes used in sequence. These include the diversification generation, the improvement stage, the updating of the refinement set, the generation of the subset and the way the solutions are combined. One way would be to construct a reference set which is based on the k promising solutions (say approximately 10 % of the entire set) which are not necessarily the top k solutions. These k solutions could be made up of say half of those with high quality (top half) and the other half constructed purposely to be diverse from the first half. A measure of dissimilarity obviously needs to be adopted. For instance, for the case of the TSP, or any routing problem, the following can be used. If $X_i \in E_1$ and $X_j \in E - E_1$ where E_1 represents the set of the top half initially, $|E_1| = \frac{k}{2}$, and E the whole set of the solutions generated, the dissimilarity between solution i and solution j can be defined as $d(X_i, X_j) = 1 - \frac{2N_{ij}}{(N_i + N_j)}$ where N_{ij} represents the number of common features (arcs) in both solutions i and j, and N_i the number of arcs in solution i. The arc with the maximum distance, say $\tilde{i}$, is selected and added to the list E_1. In other terms, $\tilde{i} = \text{Arg}\big\{ \text{Max}_i \; \underset{j}{\text{Min}} \, d\big(X_i, X_j\big) \big\}$ and $E_1 = E_1 \cup \big\{\tilde{i}\big\}$. This process is updated and continued until the other half is completed (i.e., $\big|E_1\big| = k$). There obviously could be other ways of defining the dissimilarity measure depending on a particular problem. As in GA, to avoid early convergence, one way would be to use α % of the k solutions from the high-quality solutions including the overall best and the remaining as diverse as possible. A mechanism is then used to create a subset from these k solutions, say all subsets of size 2 (i.e., pairs). Once this operation is performed, a scheme to combine those subsets is then adopted. The construction of these small clusters and the way some are combined is challenging and can be critical to the success of such an approach. These new solutions are then improved. One interesting approach, for the case of vehicle routing, is proposed by Tarantilis and Kiranoudis (2002). The idea is to select a pool of sequences of nodes known as bones, and then the promising bones are combined to generate feasible and better solutions. The way the sequences are combined in SS shares some similarities with harmony search (HS). The main steps of the SS heuristic are given in Algorithm 4.2 but for more information, see Marti et al. (2006).

Algorithm 4.2: A Basic Scatter Search Heuristic

Step 1   Generate a large set of diverse and promising solutions, say P_{size}.
Step 2   Create a reference subset (say $0.1P_{size}$) of a mixture of best solutions and diverse ones, say half each initially.
Step 3   Construct a set of new subsets which can be made of all pairs say.
Step 4   Apply a combination technique to construct a new complete solution and improve it to generate a few other trial solutions.
Step 5   Update the reference set based on Step 2.
Step 6   If certain stopping criteria are met stop; otherwise go to Step 3.

Some Thoughts

The balance between the number of elite solutions and the diverse ones is a delicate issue that needs careful exploration. This number does not need to be constant throughout the search, set a priory but adjusted using iteration number or some rules based on solution quality. For instance, the use of a range $[a, b]$ with a and b being $\frac{k}{2}$ and k, for example, can also be useful to stop one solution dominating the other. One could start with an equal number $(k/2)$ then the number of elite solutions could gradually increase, or at each iteration, k is chosen randomly in the range. The selection of the new subsets and their respective sizes could also be investigated further. The use of an aggressive combination procedure, once in a while in Step 4, to further intensify the search could also be worthwhile considering.

4.3   Harmony Search

HS was originally proposed by Geem et al. (2001) to imitate the success of music players in their improvisation when searching for a music harmony that is pleasing to hear and to appreciate. Such a harmony is made up of a combination of sounds at certain pitches played by different instruments. The aim for the musician is to identify the best pitch for each sound so when combined make up an excellent harmony. The analogy with optimisation can be seen as follows:

A given harmony relates to a given solution configuration, the sounds of the instruments represent the decision variables, their respective pitches are the values of the decision variables, excellent harmony that could not be improved any longer is a global solution, the quality of the harmony is the objective function value. Each practice by the musician(s) represents the iteration or the generation number. At each practice, the musician tries to identify new pitches based on the ones he/she remembers to be of good quality, known as the HS memory (HM), while introducing some extra changes to create a new harmony. This process is repeated until the best harmony already discovered can no longer be improved. HS is a population based approach where the HS memory contains the pool of harmonies, similar to the population of chromosomes in GA. However, the way the attributes of a new harmony (new solution) are constructed does not depend on two parents only as in GA but on all previously found harmonies in HM. These are selected with a probability known as harmony memory consideration rate (HMCR) denoted here by p_M for simplicity, otherwise the new attribute is generated randomly instead. The obtained solution (harmony) is then adjusted for possible improvement with a certain probability pitch adjusting rate (PAR) which is denoted here by p_a. This is achieved by adjusting those attributes from HM that have a certain width, say W. If the new solution is better than the worst harmony in HM, (the one with the worst fitness value or objective function value), this worst one is removed using the survival of the fittest as in GA, otherwise a new harmony is generated again. This process is repeated until there is no improvement after a certain number of trials or when the maximum number of trials is reached.

Illustrative Example

As this metaheuristic is relatively less popular, let us consider the following 3 variables non-linear optimisation problem.

$$\text{Min} F(X) = (x_1 - 1)^2 + (x_2 - 2)^2 + (x_3 - 1)^2 + 10$$

$$L_j \leq x_j \leq U_j \text{with } L_j = -5 \text{ and } U_j = 5; j = 1, 2, 3$$

For illustration purposes let $p_M = 0.90$; $p_a = 0.15$ and $W = 2$ and assume the initial HM be made up of the following three solutions (harmonies):

$X_1 = (1, 3, 2)$; $X_2 = (3, 2, 2)$ and $X_3 = (2, 1, 1)$ with their corresponding $F(.)$ values.

These solutions are ranked based on their $F(.)$ values and stored in HM as follows: $HM = \begin{pmatrix} x_1 & x_2 & x_3 & F(X) \\ 1 & 3 & 2 & 12 \\ 3 & 2 & 2 & 15 \\ 2 & 1 & 1 & 16 \end{pmatrix}$

A new solution $X' = (x'_1, x'_2, x'_3)$ is then constructed as follows:

For each attribute, x'_j let us generate a random number $\alpha \in [0, 1]$.
For instance, for x'_1, we have $\alpha = 0.70 < p_M = 0.90$ which means that x'_1 is chosen from the already existing set of values of x_1 which is $\{1, 3, 2\}$, say $x'_1 = 1$. For x'_2 let $\alpha = 0.92 > p_M = 0.90$ leading to x'_2 being chosen randomly in the range $[-5, 5]$, say, and let $x'_3 = 2$ being chosen similarly to x'_1. The new solution is then $X' = (1, 4, 2)$ and its $F(X') = F(1, 4, 2) = 15$.

The obtained solution is then tested for possible improvement. This is achieved by adjusting those chosen solutions from HM. Here, $x'_1 = 1$ and $x'_3 = 2$. For $x'_1 = 1$, let a random number in $(0,1)$ $\alpha = 0.60 > p_a = 0.15$ so $x'_1 = 1$ is unchanged, and let for $x'_3 = 2$, $\alpha = 0.10 < p_a = 0.15$ so $x'_3 = 2$ is adjusted as follows.

$$x'_3 = x'_3 + \alpha.W = 2 + 0.5 \times 2 = 3 \text{ with } \alpha = 0.5 \in [-1, 1].$$

The new solution is then $X'' = (1, 3, 2)$ and its $F\left(X''\right) = F(1, 3, 2) = 12$.

Let $X' = \text{ArgMin}\left(F\left(X'\right) = 15, F\left(X''\right) = 12\right)$ leading to $X' = (1, 4, 2)$ being the new solution to be checked against the worst solution in HM which is X_3. As $F\left(X'\right) = F(1, 3, 2) = 12 < F(X_3) = 16$; X_3 which has the worst value is then replaced by the new $X' = (1, 3, 2)$. If $F\left(X'\right) \geq F(X_3)$ there will be no change as X' is not chosen and a new solution will be generated again.

The basic HS algorithm for the following non-linear problem, say Min $F(X)$ subject to $L_j \leq x_j \leq U_j; j = 1, \ldots, n$, is given in Algorithm 4.3.

Algorithm 4.3: Basic Harmony Search Heuristic

Step 1   Initialise the harmony memory *HM* and its size, and set the parameters p_M, p_a and W.
Let $X = (X_i)_{i=1,\ldots,M}$ with $X_i = (x_i^1, \ldots, x_i^n)$ put in ascending order of $F(.)$.

Step 2   Improvise a new harmony (solution) X' using the following:
For each $j = 1, \ldots, n$

(a) Generate a random $\alpha \in [0,1]$.
(b) If $\alpha \leq p_M$, select x_j' randomly from $\{x_j^1, \ldots, x_j^M\}$ in HM and generate a random value $\beta \in [0, 1]$. If $\beta \leq p_a$, then adjust x_j' by setting $x_j' = x_j' \pm r.W$ with $r \in [0, 1]$.
(c) Else set $x_j' = L_j + \alpha.(U_j - L_j)$ with $\alpha \in [0, 1]$

Step 3   If the new solution is better than the worst in HM (i.e., $F(X') < F(X_M)$), set $HM = HM - X_M \cup X'$ and rank HM again in ascending order of $F(.)$.

Step 4   If the maximum number of trials is achieved or the maximum number of trials since the last improvement is met stop otherwise go to Step 2.

Discrete Case

For the discrete case, the feasible region is limited to a set of a finite set of combinations. For instance, in the case of the TSP with n cities, $X_i = \left(x_i^1, \ldots, x_i^j, \ldots, x_i^n, x_i^{n+1} \right); i = 1, \ldots, m$ represents the ith TSP configuration among the m chosen ones and x_i^j defines the jth city to be visited in the ith TSP configuration, and $x_i^1 = x_i^{n+1}$. The objective function is defined by the total distance travelled which is $F(X_i) = \sum_{j=1}^{n} d\left(x_i^j, x_i^{j+1} \right)$.

One way of implementing HS can be the following: In Step 1, m tours are generated either randomly or via constructive heuristics such as the nearest neighbour. In Step 2(b), x'_j can be obtained from the set $\{x^1_j, \ldots, x^m_j\}$ as long as it was not already selected where the process is repeated. In addition, if $\beta \leq p_a$, the new adjusted x'_j can then be randomly generated in its neighbourhood say $x'_j \in Nx'_j$ with

$$N\left(x'_j\right) = \left\{x_j; j = 1, \ldots, n \middle/ d\left(x'_j, x_j\right) < W\right\}.$$

Here, W can be set as $W = \eta \cdot \dfrac{\displaystyle\sum_{i=1}^{n}\sum_{j=1}^{n} d\left(x_i, x_j\right)}{n(n-1)}$; $0 < \eta \leq 1$ with the parameter η which is used to control the number of neighbouring cities. Note that $N(x_j)$ can also be defined simply by its k (say 3) nearest cities to x_j. In Step 2(c) the new city is generated randomly as long as it is not already chosen.

HS relies on its parameter values that are set at the outset. The first attempt to enhance HS was made by Mahdavi et al. (2007) who expressed the pitch adjusted ratio (p_a) and the width of adjustment (W) using an increasing linear function of the iteration number for the former and a negative exponential for the latter. A further enhancement was proposed by Pan et al. (2010) who restricted the adjustment to substituting each attribute that is considered admissible (passed both the probability tests of p_M and p_a) to the corresponding attribute of the best solution. This simple rule eliminates the estimation of the parameter W while it guides the search towards the overall best. The results obtained are very competitive while requiring less computational effort.

In the discrete case, such as the example of the TSP given earlier, Step 2(b) can be modified to accommodate those cities that have higher chance of being selected in their respective positions. For instance, instead to have the loop $j = 1, \ldots, n$, we first rank the positions according to their frequency of occurrence out of m, the city with the highest value, or one of those cities with high values, can be considered first and this process is then repeated until all cities are chosen. This selection will reduce the risk of selecting a city earlier to its relatively less-attractive position.

HS was shown to be successful in solving both constrained and unconstrained non-linear optimisation problems with complex multi-modular objective functions that are commonly used in the literature. For more details on HS and references therein, see Mahdavi et al. (2007).

Some Thoughts

The dynamic updating of the parameters including p_M, which does not seem to be tackled so far, could be explored further to make the search more adaptive. One way would be link all these three parameters not only to the iteration number but also to (i) a block of successive iterations, and (ii) the improvement in the solution quality over that block of iterations. For diversification purposes, HS may needs to accommodate some form of resetting of these three parameters as successfully implemented in SA.

HS as any other evolutionary approach is successful at the exploration stage but is relatively less effective at the exploitation stage where intensification is required. This weakness can be overcome by introducing a local search once in a while to all or to a subset of those solutions in HM making the overall approach behaving as a simple composite heuristic or even a powerful hybrid which I shall discuss in (Chap. 5). In constrained problems the generation of the new attributes could be a challenge as either feasibility is imposed and needs to be checked throughout the search or controlled infeasibility is allowed with corresponding penalties attached in the objective function.

4.4 Differential Evolution

Differential evolution (DE) is an evolutionary approach originally given in 1995 as a working paper for tackling global (continuous) optimisation problems, with the published article being in Storn and Price (1997). DE is popular especially in the area of engineering where most optimisation problems happen to be non-linear. This is mainly due to its simplicity, suitability and ease of implementation. This metaheuristic is based on the idea that a new solution can be obtained by combining attributes of previously found ones through the use of vector differences with some perturbations and randomness attached. An interesting review and references therein can be found in Neri and Tirronen (2010).

Consider an $n-$ dimensional minimisation problem with $F(.)$ as its objective function. As DE is a population-based approach, let $P(t)$ be the population retained at generation t made up of $|P(t)|$ solutions $(X_i(t); i = 1, \ldots, |P(t)|$, with $X_i(t) = (X)_{ij}(t); j = 1, \ldots, n)$. For simplicity, we refer to P as the current population. DE is based on adopting the following steps:

(i) Three distinct solutions X^0, X^1 and X^2 are pseudo-randomly chosen from the population. Some of these can include to the best solution so far or the solution at the current generation (i.e., $X_i; i = 1, \ldots, |P|$).

(ii) A trial offspring is generated by the following simple mutation scheme as $Y = X^0 + \alpha(X^1 - X^2)$ with the scale factor $\alpha \in [0, 1]$ used to control the exploration. A large value will aid exploration, whereas a small one will encourage exploitation. This trial solution (i.e., $Y = Y_j; j = 1, \ldots, n$) is also known as the mutant vector or the donor vector solution.

(iii) The exchange of genes between the donor vector (Y) and the ith current solution (X_i) is performed through the following binary crossover operator with a crossover rate (β). This is set as follows:
$$\widetilde{Y}_j = \begin{cases} X_{ij} & \text{if } u \in U(0, 1) < \beta \\ Y_j & \text{otherwise} \end{cases} ; \; j = 1, \ldots, n \text{ with } u \text{ being a}$$
random number chosen uniformly in the range $(0,1)$ and β is a crossover rate, usually $\beta \in [0.6, 1]$.

(iv) The evaluation of $F(\widetilde{Y})$ is then performed and the selection of the offspring is derived by setting:
$$X_i(t) = \begin{cases} \widetilde{Y} & \text{if } F\left(\widetilde{Y}\right) \leq F(X_i(t-1)) \\ X_i(t-1) & \text{otherwise} \end{cases}$$

In other words, the best of the two solutions namely the current i^{th} solution or the newly found one is chosen (i.e.,

$$X_i(t) = \text{ArgMin} \left[F(X_i(t-1)), F\left(\tilde{Y}\right) \right]).$$

The basic steps of a DE heuristic are given in Algorithm 4.4.

Algorithm 4.4: A Basic Differential Evolution Heuristic

Step 1   Generate the initial population P with $X_i = (X_{ij}); i = 1, \ldots, |P|; j = 1$ $,\ldots, n$ and evaluate $F(X_i); i = 1, \ldots, |P|$. Set $t = 1$ (generation counter) and set $X^{best} = \text{ArgMin}(F(X_i); i = 1, \ldots, |P|)$ and $F^{best} = F(X^{best})$.

         Define the stopping criterion, the scale factor α and the crossover rate β.

Step 2   For $i = 1$ to $|P|$

         (a) Generate pseudo-randomly X^0, X^1 and X^2 from P

         (b) Compute $Y = X^0 + \alpha(X^1 - X^2)$ 'generation of the mutant vector'

         (c) For $j = 1$ to n, set $\tilde{Y}_j = \begin{cases} X_{ij} & \text{if } u \in U(0,1) < \beta \\ Y_j & \text{otherwise} \end{cases}$

Step 3   Evaluate $F(\tilde{Y})$.

         If $F\left(\tilde{Y}\right) \leq F(X_i)$,

         set $X_i = \tilde{Y}$ and update the population P.

         If $F(X_i) < F(X^{best})$, set $X^{best} = X_i$ and $F^{best} = F(X_i)$.

Step 4   If stopping criterion not met, set $t = t + 1$ and go to Step 2 else choose $X^* = X^{best}$; $F^* = F^{best}$ and stop.

There are several items that are worth highlighting.

(i) *Mutation strategy* The mutation strategy adopted in Step 2b is the original and the simplest one. There are now several variations that are commonly used. One way is to consider the impact of the best solution so far by replacing X^0 with X^{best} leading to the new mutation strategy $Y = X^{best} + \alpha(X^1 - X^2)$. If both the best so far and the current solutions are introduced, the following rule can be extended to yield the following $Y = X_i + \alpha_1(X^{best} - X_i) + \alpha_2(X^1 - X^2)$ with α_1 and α_2 acting in a similar way as α, setting $\alpha_1 = \alpha_2$ is usually adopted. This strategy combines some form of control as it refers to the direction $\left(X^{\text{best}} - X_i\right)$ while retaining randomness through the direction given by the third part $\left(X^1 - X^2\right)$. An interesting strategy that extends the above rule is to integrate global and local neighbourhood. A possible setting is to use

$$Y = wG_i + (1 - w)L_i; w \in [0, 1]$$

with $G_i = X_i + \alpha_1\left(X^{\text{best}} - X_i\right) + \alpha_2\left(X^1 - X^2\right)$ and $L_i = X_i$ $+\alpha_3\left(X^{\text{Nbest}} - X_i\right) + \alpha_4\left(X^3 - X^4\right)$. Here X^{best} and X^{Nbest} refer to the global best and the best within the neighbourhood of X_i, respectively. Similarly (X^1, X^2) and (X^3, X^4) are randomly generated from the entire population and from the neighbourhood of X_i, respectively. The weight w reflects the importance of the guided direction $\left(X^{\text{best}} - X_i\right)$ against the random-based one. This can be a non-decreasing function of the generation or iteration number as the search at the end needs to be focussed mainly on exploitation, and hence on the solutions around the global best solution. Other developments and strategies can be found in Das et al. (2009) and in the review by Neri and Tirronen (2010).

(ii) *Scalar factor* (α) It is worth noting that the determination of the mutant vector Y shares some similarities with the way a new point in non-linear optimisation is obtained (i.e., $X(t + 1) = X(t) + \lambda.\vec{S}$ with λ denoting the step size and S the direction of movement which can be gradient-based or any other form of descent, see Fletcher (1987)). Here the direction S is based on the difference between two vector solutions instead to provide randomness. The scale factor which acts as a step size is used here to balance exploration with exploitation.

(a) It is therefore important to adopt a strategy which favours larger values initially so to explore the search space more widely and then gradually decreasing it as the search progresses, and hence more exploitation is performed. This reasoning is sensible as long as the direction also plays its part otherwise the search could reject a lot of solutions in Step 3 of the algorithm. One way would be to define $\alpha(t)$ as a non-increasing function of t. For example, a convex function with $\alpha(1) = \alpha_{max} = 0.85$ and $\alpha(t) = \alpha_{min} = 0.3 \ \forall t \geq T_{max}$. Similarly, a stepwise decreasing function with N_s partitions of the total time (or generations) T_{max}, each with $\Delta = \frac{T_{max}}{N_s}$ could be worth attempting.

$$\alpha(t) = \alpha_{min} - (k-1)\frac{\alpha_{max} - \alpha_{min}}{N_s} \ \forall t \in [T_{k-1}, T_k] \ ; \quad k = 1, \ldots, N_s$$

with $T_0 = 0$ and $T_k = k\Delta$.

(b) A good approximation of the optimal step size α can also be found by solving the following one-dimensional non-linear optimisation problem:

$$\text{Min}\,(g(\alpha)); \alpha \in [a, b] = [0, 1] \text{ with } g(\alpha) = F\left(\widetilde{Y}(\alpha)\right).$$

This can be achieved using classical numerical methods such as the bisection method, Golden section, Fibonacci search, or just a descent method based on exploration search. These do not require the differentiation of g neither its analytical solution (i.e., $g'(\alpha) = 0$). These are direct approaches based on elimination techniques with the aim to reduce the range $[a, b]$ until it is small enough. As an illustration consider Golden section search which uses the Golden section proportion $(1/\gamma) = 0.618$ and $(1/\gamma)^2 = 0.382$ and an approximation accuracy of ε (the magnitude of the last range). At each iteration, two values α_1 and α_2 are generated within the current range (initially $[a, b] = [0, 1]$ say) as follows $\alpha_1 = a + (1/\gamma)(b - a)$ and $\alpha_2 = a + (1/\gamma)^2(b - a)$. If $g(\alpha_1) < g(\alpha_2)$, set $b = \alpha_2$ and if $g(\alpha_1) > g(\alpha_2)$, set $a = \alpha_1$, both leading to a smaller range (a, b). However, if there is a tie (i.e., $g(\alpha_1) = g(\alpha_2)$) set $a = \alpha_1$ and $b = \alpha_2$ leading to an even smaller range (a, b). This process is repeated until $b - a < \varepsilon$ where the approximate solution of the scale factor is set as $\hat{\alpha} = a + \frac{(b-a)}{2}$. This process requires $\left[1 + \frac{\text{Log}\left(\frac{2\varepsilon}{\Delta_0}\right)}{\text{Log}\left(\frac{1}{\gamma}\right)}\right]$ iterations to guarantee a solution with ε (say 0.01) accuracy given Δ_0 the magnitude of the original range $[a, b]$. Golden section has also the advantage that its convergence is linear as $\lim_{n \to \infty}\left(\frac{\Delta_n}{\Delta_{n-1}}\right) = 1/\gamma$. For more details on these direct methods, see any textbook in numerical optimisation methods.

(iii) *Crossover rate* (β): The crossover rate has a direct effect on generating diverse solutions. Initially, as the aim is to explore the search area, (β) can be relatively closer to zero, say in [0, 0.50] but as the search progresses more guidance is required, and hence a larger value could be more suitable, say in the range [0.6, 0.90]. Some recent and interesting development of DE including its applications and variants can be found in the review paper by Neri and Tirronen (2010).

Some Thoughts

The updating of α and β using adaptive searches alongside the definition of the mutant vector Y are the critical key elements in furthering the success of DE. If the updating α is performed at each generation and for each individual solution, this can be too time consuming, especially if numerical method technique though yield near optimal values are adopted. One way would be to use these techniques once in a while and to concentrate on those solutions that display some interesting features.

The use of retaining randomness while controlling the search is a promising idea which can be translated into generating respective ranges that are dynamically or continuously updated with the aim to have the latest ones small enough and closer to unity. A similar scheme is also highlighted in Chap. 3 regarding the update of the temperature in SA.

The mutation strategy adopted is probably the most challenging but the one which can have a wider scope for improvement. For instance, an integration of neighbourhood information and global information while retaining some form of randomness in the mutation strategy, though some attempts were already made, could be revisited and extended further. One of the questions is how to define the neighbourhood and how big it is so it does not ignore important features of the search while being unnecessary big. As DE is an evolutionary-based technique some form of hybridisation with other local searches or metaheuristics that tend to exploit the search better can also be worth exploring.

4.5 Ant Colonies

The ant system (AS) is a metaheuristic inspired by the observation of the behaviour of real life ant colonies, in particular, the way in which real ants find the shortest path between food sources and their nest. Good quality solutions are developed as a result of the collective behaviour of the ants which is achieved via a form of indirect communication obtained by depositing a substance called pheromone on to the ground which forms a pheromone trail. The effect of this is that the pheromone trail will build up at a faster rate on the shorter path. This will influence more ants to follow the shorter path due to the fact that the ants prefer to follow a path with a higher pheromone concentration. As a greater number of ants chooses the shorter path, this in turn causes a greater level of pheromone, and hence encourages more ants to follow the shorter path. In time, all ants will have the tendency to choose the shorter path. In other words, the greater the concentration of the pheromone on the ground is, the higher the probability that an ant will choose that path. For more information, see Daneubourg et al. (1990).

This real life behaviour of ants has been adapted to solve combinatorial optimisation problems. AS algorithms employ a set of agents, known as ants, who search in parallel for good solutions using a form of indirect communication. The artificial ants co-operate via the artificial pheromone level deposited on arcs which is calculated as a function of the quality of the solution found. The amount of pheromone an ant deposits is proportional to the quality of the solution generated by that ant helping direct the search towards good solutions. The artificial ants construct solutions iteratively by adding a new node to a partial solution exploiting information gained from past performance using both pheromone levels and a greedy heuristic.

Introduction to Ant System

The first ant colony optimisation algorithm was initially proposed by Colorni et al. (1991) to solve the TSP. Originally, three algorithms, ant-cycle, ant-density and ant-quality, were put forward. Ant-density and ant-quality updated the pheromone level on an edge as soon as that edge had been selected by an ant, while in ant-cycle, the pheromone was only updated after all ants had constructed their tours. Ant-cycle was found to perform better than the other two which was then adopted as the AS. This variant uses artificial pheromone trail values, τ_{ij}, associated with each arc (i, j). Initially, m ants are placed on randomly selected nodes and each ant starts constructing a tour from this starting position. The tour is built by each ant successively choosing the next customer to visit probabilistically using the following selection rule.

Selection Rule

The probability that an ant currently situated at customer i will visit customer j next is given by the state transition rule in Eq. (4.1):

$$
p_{ij} = \begin{cases} \dfrac{\tau_{ij}^{\alpha} \cdot \eta_{ij}^{\beta}}{\displaystyle\sum_{l \in F_i^k} \tau_{il}^{\alpha} \cdot \eta_{il}^{\beta}} & \text{if } j \in F_i^k \\[2ex] 0 & \text{otherwise} \end{cases} \tag{4.1}
$$

where

η_{ij} is a local heuristic function, known as the visibility, usually given as $\eta_{ij} = 1/d_{ij}$.

d_{ij} is the Euclidean distance from customer i to customer j.

F_i^k is the list of feasible customers not yet visited by ant k.

The parameters α and β relate to the relative influence of the trail value and the heuristic information, respectively. If α is set too high compared to β, then the solutions will stagnate producing poor solutions, but if α is set too low compared to β, then the algorithm behaves as a stochastic greedy heuristic due to the fact that ants are initially randomly placed on nodes. A balance has to be struck between α and β. For instance, if α is set to zero, then the transition rule becomes a greedy algorithm choosing the closest customers, however if β is set to zero, then only the pheromone trail value is taken into account which will cause stagnation in the solutions.

Local Updating

Once all ants have constructed their respective complete tours, then the pheromone trail levels are updated by each ant according to the updating rule given in Eq. (4.2). Here all arcs have their pheromone reduced due to evaporation but at the same time those arcs that belong to the tours generated by the ants are given extra pheromone, which is inversely proportional to the length of the tours. This is introduced to encourage those chosen arcs to be more attractive for selection in the future.

$$\tau_{ij} \leftarrow (1 - \rho)\tau_{ij} + \Delta_{ij} \tag{4.2}$$

Where
$0 < \rho < 1$ is the pheromone trail evaporation.
$\Delta_{ij} = \sum_{k=1}^{m} \Delta\tau_{ij}^{k}$ where the quantity $\Delta\tau_{ij}^{k}$ is set as follows:

$$\Delta\tau_{ij}^{k} = \begin{cases} \dfrac{Q}{L_k} & \text{if } (i,j) \in T^k \\ 0 & \text{otherwise} \end{cases} \tag{4.3}$$

with T^k being the tour produced by ant k, Q a correction factor used to normalise the data set and $L^k(t)$ the length of the tour T^k at time (iteration) t.

Note that the parameter ρ is used to lower the pheromone trail on all arcs to avoid stagnation in the solution. This setting will stop some trails to accumulate an excessive amount of pheromone which could lead to a number of those arcs dominating the solutions and leading to early convergence and poor local minima. $\Delta \tau_{ij}^k$ is the amount used to reinforce trail values on those arcs that are visited by ants.

Global Updating

Here, an elitism strategy that provides a strong reinforcement to the tour corresponding to the global best solution is used. Arcs belonging to the global best tour have their pheromone increased by a quantity e. L^{gb} with e representing the number of ants that produce the best tour and L^{gb} being the length of the global best tour. The consequence of the global updating rule is that arcs belonging to short tours and arcs which have been used frequently are favoured by receiving a greater amount of pheromone.

The AS Algorithm

The main steps of the AS heuristic are given in Algorithm 4.5.

Algorithm 4.5: A Basic AS Heuristic

Step 1   Initialise trail values and placement of say M ants.

Step 2   While none of the stopping criteria is met (say $iter < K$), do the following:

(i) For each ant ($m = 1,...,M$), given their initial starting point, perform the following tasks:

- Select the next customer to visit according to the state transition rule using Eq. (4.1).
- Execute local updating of the pheromone using the trail updating rule using Eq. (4.2).

(ii) Apply local search to refine the solutions (*optional step*)

Step 3   Apply global updating to reinforce the best solutions.

ASs have been used successfully to solve many hard combinatorial problems including the TSP, the sequential ordering problem, the quadratic assignment problem, the vehicle routing problem, among others.

Ant Colony System

The Ant Colony System (ACS) was proposed by Dorigo and Gambardella (1997) to improve AS by introducing the following three modifications: (a) a different transition rule is applied, (b) the pheromone global updating rule is based only on the global best solution, and (c) each time an ant selects an arc then the pheromone level on that arc is reduced, known as the local updating rule.

Modified Transition Rule

In ACS, ants choose the next node, j, to visit by means of the pseudo-random-proportional rule given in Eq. (4.4).

$$j = \begin{cases} \text{argmax}_{u \in F_i^k} \tau_{iu} \eta_{iu}^\beta & \text{if} \quad q \leq q_0 \\ S & \text{otherwise} \end{cases} \tag{4.4}$$

Where q is a random number uniformly distributed in $(0, 1)$, q_0 a parameter $(0 \leq q_0 \leq 1)$ and S a random variable chosen based on the probability distribution (state transition rule) given earlier in Eq. (4.1). It can be noted that this transition rule reduces to the one used by AS when $q_0 = 0$ or when $q > q_0$. In other words, If $q \leq q_0$, then the best arc is chosen, hence encouraging exploitation of the available knowledge; otherwise, the edge is chosen as before encouraging biased exploration. The value of q_0 can be critical to the success of the algorithm. It is usually set closer to 1 (say 0.85) though it is more appropriate if such a value is adaptively updated depending on the quality of the solution rather than fixed a priory. The higher the value of q_0 is, the more the method behaves as a pure descent method. One way would be to start with 0.50, then slowly decreasing it or increasing it based on the quality of the obtained solutions.

In ACS, only the ants belonging to the global best tour are used to update the level of pheromone. The global updating rule is similar to the standard one except that here both the pheromone value and the heuristic visibility are linearly weighted, see Eq. (4.5).

$$\tau_{ij} \leftarrow (1 - \rho)\tau_{ij} + \rho\Delta\tau_{ij} \qquad (4.5)$$

$$\text{and } \Delta\tau_{ij} = \begin{cases} 1/L^{\text{best}} & \text{if } (i,j) \in \text{best tour} \\ 0 & \text{otherwise} \end{cases}$$

Note that the best tour can be represented by the global best or the iteration best. The global update rule allows only the arcs belonging to the best tour to be adjusted so that the trail value is increased according to the quality of the solution.

Local Updating

The local update rule introduced in the ACS method is applied to an arc as soon as it has been selected by an ant buildings its tour. If an ant currently situated at customer i chooses customer j to visit next, then the pheromone level on arc (i, j) is updated by the local update rule given by Eq. (4.6).

$$\tau_{ij} \leftarrow (1 - \nu)\tau_{ij} + \nu\Delta\tau_0 \qquad (4.6)$$

where ν is the a pheromone decay parameter chosen in the interval $(0, 1)$, τ_0 is the initial pheromone trail level set at the beginning of the algorithm usually set to $\tau_0 = \frac{n}{L_H}$ where n is the total number of customers and L_H is the length of a tour produced by a given heuristic.

The local updating rule is designed to encourage exploration of not yet visited arcs. The trail value on an arc already chosen is reduced in order to make it less attractive to successive ants. The rule may also encourage arcs visited early in one tour to be visited later in subsequent tours.

Other Variants

Stützle and Hoos (2000) introduced the Max-Min AS (MMAS). Their method is basically similar to AS using the same state transition rule in the selection process with some modifications, the main one is to restrict the pheromone trail values to be in a certain interval $[\tau_{\min}, \tau_{\max}]$. The aim is to reduce the risk of stagnation as highlighted earlier. Though the idea of introducing restriction on the trail values is interesting as it avoids stagnation and provides a chance for less-visited solutions to exist, the use of these additional parameters τ_{min} and τ_{max} need to be defined appropriately.

Bullnheimer et al. (1998) introduced another modification to AS by ranking the ants according to their tour length and the amount of pheromone an ant contributes. At each iteration, only the $(\sigma - 1)$ best ants are used to deposit the pheromone. The global best solution is then updated according to the weight σ, whereas those arcs belonging to the tour of the μth best ant are updated according to the weight $(\sigma - \mu)$, where μ is the ants rank. The global update rule is defined as follows:

$$\tau_{ij} \leftarrow (1 - \rho)\tau_{ij} + \sum_{\mu=1}^{\sigma} (\sigma - \tau + 1)\Delta\tau_{ij}^{\mu} \qquad (4.7)$$

where

$$\Delta\tau_{ij}^{\mu} = \begin{cases} \dfrac{Q}{L^{\mu}} & \text{if } (i,j) \in \text{a tour of and with rank } \mu, \ \mu = 1, \ldots, \sigma \\ 0 & \text{otherwise} \end{cases}$$

and L^{μ} is the solution cost of the tour with rank μ. If $\mu = \tau = 1$, the global best solution is used only. This is similar to the original updating rule and hence Eq. (4.7) reduces to Eq. (4.5). The parameter Q is a correction factor used to normalise the data. It has been shown that the AS_{Rank} procedure generally produced significantly better results than AS, see Dorigo et al. (1999).

Wade and Salhi (2003) investigated the use of Ant Colony Optimisation (ACO) in a class of VRP with mixed pickups and deliveries where the backhauls can occur anywhere in a delivery route. They introduced some variations by catering for the characteristics of the problem. For simplicity, we present the selection rule, and both the local and the global updating rules.

Selection Rule The next node to visit after i is selected based on the following state transition rule

$$j = \begin{cases} \text{ArgMax}_{u \in F_i^k} \left\{ \tau_{iu}^{k}{}^{iu_{iu}} \eta_{iu}^{\beta} . \kappa_{iu}^{\lambda} \right\} & \text{if } q \leq q_0 \\ S & \text{otherwise} \end{cases}$$

where both q and q_0 are defined as earlier and S is selected according to the density function

$$p_{ij} = \frac{\tau_{ij}^{\alpha} . \eta_{ij}^{\beta} . \kappa_{ij}^{\lambda}}{\sum\limits_{\mu \in F_i^k} \tau_{iu}^{\alpha} . \eta_{iu}^{\beta} . \kappa_{iu}^{\lambda}}$$

Here, visibility is represented by $\eta_{ij} . \kappa_{ij}$ where these two factors are expressed depending on the remaining load of the vehicle (spare capacity) after visiting node i and the nearest neighbour to node i.

Local Updating Once an arc is chosen by a given ant, its trail is updated while considering the frequency of occurrence n_{ij} of arc (i, j). Such an arc is penalised only if it happens to be chosen more than a certain number of time say $\bar{n}$. This frequency-based local updating is given as follows:

If $n_{ij} > \bar{n}$ then $\tau_{ij} \leftarrow (1 - \nu\gamma)\tau_{ij} + \nu\tau_0$

Else use Eq. (4.6) with $\gamma = 1$

Where γ refers to the correction factor and all other notations are defined earlier.

Note that $\bar{n}$ needs to be carefully set as a balance between exploration and exploitation has to be struck, with $\bar{n} = 0.8$ was found to be suitable.

Global Updating Elitism and ranking are both combined to yield a compromise global updating rule. Here, instead of referring to the maximum number of elite solutions allowed for the global updating of the trails, a restriction that considers the solution quality is also introduced into the selection rule. This restriction shows that only those solutions that lie within θ % of the global best are potential candidates, not all the best solutions as commonly used.

For more information, references and applications on ASs in general,, the review paper by Dorigi and Stutzle (2010) can be a useful reference.

Some Thoughts

The success of ACS relies on the way the problem is formulated and represented, how visibility is expressed within the state transition rule, how the ants are placed, the construction of the candidate list (e.g., site-dependent), and how the elitism and ranking are combined within the global updating rule. Some of these issues were investigated by several authors but remain open for investigation. The hybridisation of ACS with other powerful local searches is also worth exploring.

4.6   The Bees Algorithm

This optimisation algorithm is inspired by the behaviour of swarms of honey bees when searching for their food. It shares some similarities with the way the ants behave. The best example of individual insects to resolve complex tasks is by the collection and processing of nectar. Each bee decides to reach the nectar source by following another bee which has already discovered a patch of flowers. The communication between bees is based in the Waggle dance of those bees that have previously found a rich food source. Remarkably, via this informative dance, the bee has actually informed its hive mates about the direction and distance of the food source. The direction is expressed via the angle of the dance relative to the sun position, whereas the distance is expressed via the length of the straight waggle run.

Once the bee returns to the hive with some food, the bee can (a) abandon the food source and then become again an uncommitted follower, (b) continue to forage at the food source without recruiting new bees, or (c) dance and thus recruit more bees before it returns to the food source.

The basic bee algorithm originally developed for continuous optimisation problems is given in Algorithm 4.6 but more details can be found in Pham et al. (2006). In brief, the bees are first assigned to some chosen sites with the idea that those sites that attract more bees are considered more promising. This is repeated several times till a certain stopping criterion is met. The way a site is considered attractive is the most important point. This is measured by the fitness function of each bee at that site where the top say K sites will receive more bees either in an equal number or proportionally to their fitness.

Algorithm 4.6: A Basic Bees Algorithm

Step 1   Initialise the population of bees with random solutions and evaluate their fitness.

Step 2   While stopping criteria not satisfied, do the following:

2.1 Select sites, say N, and recruit bees for these sites by assigning more bees to the best say K sites and then evaluate their fitness.

2.2 Select the fittest bee from each site.

2.3 Let the rest of the bees explore randomly the neighbourhood and evaluate their fitness.

2.4 Determine the best K sites using the best fitness values or an aggregate measure such as the average fitness values of the bees at those sites.

Another popular but related bee algorithm is the Artificial Bee Colony (ABC) algorithm proposed by Karaboga (2005). This is also designed for real-parameter optimisation. The ABC algorithm is inspired by the intelligent foraging behaviour of honeybee swarm. The foraging bees are classified into three categories: (a) employed, (b) onlookers and (c) scouts. All bees that are currently exploiting a food source are classified as 'employed'. Employed bees, the more experienced bees, are responsible for exploiting the nectar sources explored before and giving information to the waiting bees (onlooker bees) in the hive about the quality of the food source sites which they are exploiting. Onlooker bees wait in the hive and decide on a food source to exploit based on the information shared by the employed bees. Scouts either randomly search the neighbouring areas to find a new food source depending on an internal motivation or based on possible external clues. Scout bees can be visualised as performing the job of exploration, whereas employed and onlooker bees can be considered as performing the job of exploitation.

The ABC algorithm is an iterative algorithm that starts by assigning all employed bees to randomly generated food sources (solution). At each iteration, every employed bee determines a food source in the neighbourhood of its currently associated food source and evaluates its nectar amount (fitness). If this new fitness is better than the previous one, then that employed bee moves to this new food source; otherwise, it retains its old food source. When all employed bees have finished this process, the bees share the nectar information of the food sources with the onlookers, each of whom selects a food source according to the food quality.

Clearly, with this scheme good food sources will attract more onlookers than those with relatively poorer quality. After all onlookers have selected their food sources, each of them determines a food source in its neighbourhood and computes its fitness. If an onlooker finds a better fitness value, it changed the employed food source with this new information. However, if a solution represented by a particular food source does not improve for a predetermined number of iterations then that food source is abandoned by its associated employed bee which becomes a scout (i.e., it will search for a new food source randomly). The whole process is repeated till the termination condition is met.

The main steps of the ABC algorithm are given in Algorithm 4.7 but more details can be found in Szeto et al. (2011) where efficient implementations are presented for solving the vehicle routing problem.

In brief, from a combinatorial optimisation view point, the algorithm uses a combination of neighbourhood search and random search to imitate such behaviour. It works similarly to the ACS where an initial number of bees known as scout bees are first assigned (recruited) either randomly or based on certain rules to discover some promising regions (exploration). The chosen regions and their respective neighbourhoods are then selected for intensification using more bees known as follower bees. The number of bees assigned to a given region (patch) depends on the fitness value assigned to that region (e.g., the fitter bees will attract relatively more followers, etc.). The search is repeated several times and the overall best solution is chosen as the global best solution.

Algorithm 4.7: The ABC Algorithm

Step 1   Initialisation step

1.1 Generate a number of initial solutions (say P) as the initial food sources $(X_i, i = 1, \ldots, P)$ and evaluate their fitness function, say $F(X_i)$. Assign each employed bee to a food source.

1.2 Initialise T_{max} and L_{max} as the maximum number of iterations (time) and the maximum number of successive iterations without improvement, respectively. Let $L_i = 0 \forall i = 1, \ldots, P$ the current number of iterations without improvement for the ith bee and set the set of neighbours of X_i, say $N(.)$, to $S_i = \{\emptyset\}$ and $iter = 0$ (iteration count).

Step 2   While stopping conditions not met do the following:

2.1 For each food source, X_i find $X_i' \in N(X_i)$.
If $F(X_i') < F(X_i)$, set $X_i = X_i'$ and $L_i = 0$; else set $L_i = L_i + 1$

2.2 For each onlooker, select a food source X_i randomly or pseudo-randomly, find $X_i' \in N(X_i)$ and set $S_i = S_i \cup \{X_i'\}$

2.3 For each food source, X_i and $S_i \neq \emptyset$ find $X_i' = \text{ArgMin}_{X \in S_i} F(X)$
If $F(X_i') < F(X_i)$,   then   select $X_j \in F$ with $j = \text{ArgMin}_{i=1,\ldots,|F|} \{L_i \text{ st } F(X_i') < F(X_i) \text{ and } X_j \in F\}$,   and   set
$X_j = X_i'$ and $L_i = 0$
else set $L_i = L_i + 1$

2.4 For each food source X_i
If $L_i = L_{max}$, then find $X_i' \in N(X_i)$ and set $X_i = X_i'$

2.5 If $iter < T_{max}$, set $iter = iter + 1$ and go to step 2.

For other algorithms inspired by bees' behaviour, see Wedde et al. (2004), Yang (2005), and also Gomez et al. (2013).

Some Thoughts

The number of scouts and the number of employed used overall and in each patch initially and during the search is crucial. The way the fitness function is computed at a given site (singly or an average over a certain number of bees) and how the communication is passed between the scouts and the employees to decide whether to abandon completely a given patch or to reinforce other patches are also important key factors that guide the search to better solutions. The integration of ABC and other bees-based algorithms also have the weakness of being less intensive at pinpointing the good local minima, and hence their hybridisation with one point move solution methods is one way forward. This issue will be visited in the next chapter.

4.7 Particle Swarm Optimisation

This is inspired from the social behaviour of a flock of birds or fish schooling (or bees). This evolutionary stochastic heuristic is introduced by Kennedy and Eberhault (1995) where the population of individuals, say of size n, which searches the promising region is called a swarm and the individuals are referred to as particles (individual solutions). It is interesting to stress the similarities which exist in the classical non-linear numerical optimisation techniques where the next point is based on the previous point and the displacement. In other words, the new point at iteration $(k + 1)$ lies along the direction S_k from the previous point X^k with an optimal step size λ_k (i.e., $X^{(k+1)} = X^{(k)} + \lambda_k S_k$). In these numerical optimisation methods, the aim is to design a suitable direction then to derive optimally or numerically the value of the step size, see, for instance, the classical textbook in numerical optimisation by Fletcher (1987).

Here, a particle i (at position p_i) is flown with a velocity V_i through the search space, but retains in memory its best position ($\vec{p}_i$). In the global Particle Swarm Optimisation (PSO), each particle, through communication, is aware of the best position of the particles of the swarm ($\vec{p}_g$).

At a given iteration k, the position of the ith particle ($i = 1 \ldots n$) is updated using Eq. (4.8) as follows:

$$X_i^k = X_i^{(k-1)} + V_i^k; i = 1, \ldots, n \qquad (4.8)$$

where the velocity (or displacement) is defined as a linear combination of three velocities, namely, (a) the velocity at the previous iteration, (b) velocity with respect to the best position of this particle up to this iteration p_i^k, and (c) velocity with respect to the global best position of all particles up to this iteration p_g^k. This can be written as follows:

$$V_i^k = wV_i^{(k-1)} + c_1\alpha_1\left(p_i^k - X_i^k\right) + c_2\alpha_2\left(p_g^k - X_i^k\right) \qquad (4.9)$$

where $\alpha_1, \alpha_2 \in [0, 1]$ and c_1, c_2 are cognitive and social parameters, usually set to 1.5. w is the inertia weight generated in $[0,1]$, usually set to 0.4.

Initially the values of $X_i^0 \in [X_{\min}, X_{\max}]$ and $V_i^0 \in [V_{\min}, V_{\max}]$

Note that Eq. (4.9) can be simply substituted by setting $\beta_1 = c_1\alpha_1$ and $\beta_2 = c_2\alpha_2$ which are both usually randomly generated in the range [0, 1.5]. This can then be rewritten as

$$V_i^k = wV_i^{(k-1)} + \beta_1\left(p_i^k - X_i^k\right) + \beta_2\left(p_g^k - X_i^k\right) \qquad (4.10)$$

The main steps of a basic PSO are given in Algorithm 4.8.

PSO is found to be suitable for solving combinatorial optimisation and especially <u>unconstrained global optimisation</u> problems. In the case of constrained optimisation, it may not be easy to guide the search as X_i^k needs to be feasible. In some instances, even initial solutions could be difficult to generate. One way is to guide the updating through a restricted velocity based on the amount of infeasibility as shown by Lu and Chen (2006). For instance, V_i^k can be replaced by $\left(p_g^k - p_r^k\right)$ to reflect the size of the region around (p_g^k) in the feasible region, where r is randomly generated in the range $[1, n]$ and (p_r^k) is the best position of particle r at iteration k.

Algorithm 4.8: A Basic PSO Heuristic

Step 1   Initialise the population of the n particles:
For each particle, $i = 1 \ldots n$
Generate randomly $X_i^0 \in [X_{\min}, X_{\max}]$ and $V_i^0 \in [V_{\min}, V_{\max}]$. Set $k = 1$.

Step 2   Compute the corresponding objective function for each particle i and update p_i^k and p_g^k.

Step 3   Update the position X_i^k and velocity V_i^k of each particle using Eqs. (4.8) and (4.9) or (4.10) respectively.

Step 4   Set $k = k + 1$ and repeat Steps 2 and 3 until a suitable stopping criterion is met.

As the swarm may become stagnated (early convergence) after a certain number of iterations, a form of diversification or perturbation is often recommended. One way would allow at each iteration a small probability for each particle, and to have its initial position and its initial velocity reinitialised randomly. This simple chaotic perturbation adds diversity to the system and avoids the search from getting stuck. This could also be obtained by adding a weighted extra term (a noise) to the last velocity update equation. For example, the extra term can be of the form $\beta_3 \left(Y_i^k - X_i^k \right)$ with β_3 as the corresponding weight and Y_i^k denoting a random position. The higher the value of the weight, the more exploration the PSO will perform and vice versa. If the weight is set to zero, the velocity update formula reverts back to the original one, see Garcia-Villoria and Pastor (2009) for more details on certain diversity schemes for PSO. Another implementation is to have the checks for the velocity and the position to be carried out separately.

Some Thoughts

To reduce the risk of stagnation earlier in the search, a more informative implementation would be to apply a chaotic perturbation that is time-dependent by allowing a higher probability earlier in the search and relatively a smaller one later on. Note that in these approaches, there is no reference to the step size λ_k at iteration k as used in non-linear optimisation. Here, this could be seen to be part of the velocity update, however, I believe that the inclusion of such a step size, which needs to be derived probably approximately, is worth the consideration. Note that if the problem is multi-dimensional, say of dimension D, the previous variables (X_i, p_i, p_g) can be replaced by (X_{id}, p_{id}, p_{gd}) for $d = 1 \ldots D$ respectively.

PSO could also be enhanced as the other population-based methods by incorporating intensive-based searches at certain well-defined iterations that need to be identified. Given that PSO is usually geared towards solving non-linear optimisation, adopting such an approach to discrete problems is studied relatively much less, and hence an efficient transformation on how the neighbourhoods can be represented would be interesting to explore. One way would be to transform the X_i^k in Eq. (4.8) into corresponding discrete values which may not be that simple.

4.8 A Brief on Other Population-Based Approaches

In this section, other metaheuristics that also fit in this category such as path relinking, cross-entropy (CE), artificial immune system (AIS), plant propagation and psycho-clonal algorithms are briefly discussed. This is only a few though there are many others that could also be worth exploring.

Path Relinking

This is based on evolutionary concepts as in GA except that no randomness is incorporated into the search. This is introduced as a way of integrating intensification with diversification, see Glover and Laguna (1997). The idea is to systematically generate new solutions through the construction of a path (or a trajectory) between an initial solution (known as the initiating solution) and a good solution from the set of elite solutions (known as the guiding solution). This is performed by transforming, through simple moves, the initial solution into the final one by selecting a move that contains attributes from the elite solutions with the aim to produce promising intermediate solutions in the process. The list of elite solutions is updated to reflect the solution quality and the diversity within the solutions. The size of the set of the elite solutions and the choice of the guiding solution from the list can be critical to the success of this approach. One way is to choose a move that minimises the number of moves to reach the guiding solution and which may generate new good intermediate solutions that could be used as promising starting solutions later on. The generation of the elite solutions is similar to the reference set commonly used in SS. The hybridisation of path relinking with GRASP has proved to be popular and promising, though other combinations, for example, with TS and VNS also produced competitive results. For further details and references, see Resende and Ribiero (2005).

CE-Based Algorithms

CE is an iterative population-based method made up of two steps that are applied in sequence until a stopping criterion is met. These steps include (i) the generation of feasible solutions pseudo-randomly based on a probability distribution representing the frequency of the occurrence of the attributes of a given solution, and (ii) the way the distribution is updated. The idea is that this adaptive technique will have the tendency to estimate better the probabilities, and hence generate better solutions. CE was initially presented by Rubinstein (1997) for estimating rare events (financial risk, false alarms in radar, etc.) and then Rubinstein and Kroese (2004) put together a formal description of this approach and presented its uses in combinatorial optimisation and neural computational. In brief, this can be described as follows:

(a) Initialise the probabilities of all attributes (say e), P_e^k at iteration k (initially $k = 0$).
(b) Repeat the following until a stopping criterion is met.

   – Generate N feasible solutions pseudo-randomly based on the probability P_e^k
   – Select αN best solutions ($0 < \alpha \leq 1$) to form the subset S^k.

– For each attribute e in S^k, find the number of solutions that contain attribute e, say

N_e^k, and update the probabilities P_e^k as $P_e^{k+1} = \frac{N_e^k}{|S^k|}$.

The update of the probabilities can also be carried out by setting

$P_e^{k+1} = \beta P_e^k + (1 - \beta)\frac{N_e^k}{|S^k|} (0 \leq \beta \leq 1)$, or using other suitable formula.

Caserta and Quinonez Rico (2009) presented an efficient implementation of this approach for solving large capacitated location problems. Note that CE has some basic similarities with ACO in the way that the new solutions are selected using a probability distribution.

Artificial Immune Systems

The immune system's aim is to defend us against diseases and infections. It recognises antigens using immune cells which are known as B-cells whose jobs are to circulate through the blood continuously watching and waiting to encounter antigens (foreign molecules of the pathogens). Each antigen has a particular shape that can be recognised by the receptor of the B cell. AIS is emerging fast in some applied areas of computer science such as data analysis and data mining, pattern recognition and now it is adapted to the area of optimisation. Though it started to take pace only over 15 years ago when the first book by Ishida et al. (1998) was edited in Japanese, it is only a year later when Dasgupta (1999) edited a volume of selected papers in the area of immunology and AISs that gave this field a good push. The first textbook on this area was written a few years later by de Castro and Timmis (2002). There are now several applications and algorithms based on this concept. The idea is originated from human immune systems which are capable in recognising, defending and reacting to pathogens. In other words, when a pathogen invades the organism it was observed that a number of immune cells which recognise the pathogen will reproduce in large numbers, known as clones. This process is known as reinforcement learning. The clones are then diversified using two methods: (i) a high mutation rate (known as hyper-mutation) which is performed by introducing random changes, and (ii) receptor editing which aims to remove the less-attractive antibodies (poor solutions) and replaced them with new ones. In this way, those cells that bind with their antigens are multiplied, whereas the others are eliminated following the survival of the fittest. In addition, some of the successful ones are also kept in memory to face future similar invaders. This concept is similar to the intensification of the search in promising regions, whereas the hyper-mutation acts as a diversification strategy. In brief, AIS shares some similarities with GA with the exception that there is no crossover but just a hyper-mutation. An interesting mapping between optimisation and immune system for the case of the TSP is given by Bakhouya and Gaber (2007).

The Plant Propagation Algorithm

First introduced by Salhi and Fraga (2011), the plant propagation algorithm (PPA) is a nature-inspired heuristic that emulates the way plants and, in particular, the strawberry plant propagates. Unlike most evolutionary techniques that relate to animal or insect behaviour, PPA considers plants instead as highly emphasised by the Darwinian theory of species in the wider sense. It is observed that the strawberry plant has a particular way of propagating not only through seeds but also mainly through its runners. A runner is a long branch which grows over ground and only when it touches it (i.e., where the place for growth is chosen), the runner produces roots (i.e., invest its effort) which then give rise to another strawberry plant. In other words, if the runner selected a good place for growth, there would be more chance of producing a new healthy strawberry plant. The analogy with combinatorial optimisation is that the use of runners to explore the landscape where they are to find good places to grow and propagate refer to the exploration of the search space. For the plant, a good place is one which is sunny, has enough nutrients and humidity, whereas for optimisation this relates to promising neighbourhoods. The healthy plants will survive, whereas the others will die away which is the same reproduction mechanism usually used in evolutionary techniques (survival of the fittest). To improve its chances of survival in nature, a strawberry plant adopts the following strategy that incorporates two interesting factors (i) if a good spot is detected, many short runners are dispatched there but (ii) if a poor spot is identified a few long runners are used only. In other words, in (i) exploitation or intensification is used to identify local minima, whereas in (ii) some form of exploration is conducted instead acting as its diversification strategy. These two interesting properties which combine the concentration of the search by sending a lot of short runners and a wider exploration by sending a fewer long runners are also found in the behaviour of bees and ants as described earlier in this chapter.

An overview of PPA, that considers the case of a continuous minimisation problem $\{\operatorname{Min} F(X); X \in S \subset \mathfrak{R}^n\}$, where this technique was originally used for, is summarised as follows:

A population of solutions of size M is first generated with $P = (X_1, \ldots, X_M)$ and their $F(X_i); i = 1, \ldots, M$ computed and ranked. Let the per cent of plants or solutions considered to be in good places is denoted by α, the number of long runners by N_L and the number of short runners by N_S with $N_S \gg N_L$. In other words, whenever a good place is found (i.e., a promising X_i) among these α sites, we generate N_S neighbouring solutions $(X'_k \in N(X_i); k = 1, \ldots, N_S)$. However, in poor areas (i.e., not those α sites), for diversification purposes, we also generate a relatively smaller number of neighbouring solutions $(X'_l \in N(X_i); l = 1, \ldots, N_L)$. This construction leads to $M + \alpha N_L + (1 - \alpha)N_L$ number of solutions including the original ones. The top M solutions are chosen and the process is repeated until a stopping criterion is met.

It is worth noting that this selection (survival of the fittest) shares some similarities with other evolutionary techniques such as GA. The power of PPA is that the new generated solutions are guided through the concentration of the promising regions or neighbourhoods, a similar strong guiding scheme that showed to be powerful in some VNS-based variants.

The setting of the main three parameters α, N_S and N_L is usually based on the objective function value or their fitness (see Salhi and Fraga 2011) making PPA relatively more robust compared to other evolutionary techniques. PPA is shown to produce encouraging results for both continuous global optimisation problems (Salhi and Fraga 2011) and discrete optimisation problems such as the TSP (Selamoglu and Salhi 2016). This metaheuristic has also been applicable with competitive results in industrial applications as discussed in Sulaiman et al. (2014). Though PPA is still in its infancy, the way the three parameters α, N_S and N_L are updated alongside, the incorporation of additional components for hybridisation purposes would definitely make PPA even stronger and more exciting and challenging to explore. For instance, α and N_S could be defined as non-decreasing functions of the generation number, whereas N_L could follow a non-increasing function instead. Very recently, an interesting variant which emulates propagation through both runners and seeds is presented with promising results by Sulaiman and Salhi (2016).

Psycho-Clonal Algorithm

This metaheuristic was initially developed by Tiwari et al. (2005) and it is based on the AIS as discussed earlier (mainly based on the clonal selection) and the theory of hierarchy of social needs as proposed by Maslow (1954). This hierarchy is composed of five levels where the lowest Level A refers to the physiological needs (each antibody represents a solution), Level B represents safety needs (the evaluation of the objective function), Level C the social needs (the best solutions are selected and cloned proportionally to their objective function value), Level D the growth needs (diversification used to generate new solutions via hyper-mutation) and finally the highest Level E refers to the self-actualisation needs (the best solutions are chosen to be part of the new population including the injection of new ones).

As an example, consider the example of the TSP where we have the following:

In level A, a set of solutions (tours) is generated, say N, and their affinity function $F(.)$ (similar to a fitness function as in GA) is defined (say 1/tour length). In Level B, for each solution, the affinity function is evaluated. In Level C, the best N_b solutions are chosen to be cloned. The kth best selected solution will produce N^k_{clone} clones where such a mapping is non-increasing with k, reflecting the quality of the solution. In other words, the best will be rewarded by allowing it to produce more clones than the second best, and so on where the last selected one will be allowed to produce the smallest number of clones.

One way would be to use $N^k_{\text{clone}} = \left[\frac{\alpha N}{k}\right]$ where $k = 1,\ldots, N_b$ and α is a weight coefficient (usually it is set to 1.5).

In Level D, the clones are hyper-mutated with a rate inversely proportional to their affinity. A solution is then transformed by mutation through exchanging positions within the tour and the new affinity function is evaluated. In Level E, the worst solutions are dropped and replaced either randomly or using intelligent removal/insertion strategies. The whole process is repeated from Level B until a stopping criterion is met.

This approach can be considered as a mixture of AIS and guided GA where the management of the population is maintained through a guided way based on the quality of cloning, whereas diversity is controlled through the hyper-mutation.

4.9 Summary

In this chapter, some of the commonly used population-based heuristics are reviewed. Those that are mainly geared towards combinatorial or discrete optimisation include GA, SS, ant colony and bees algorithms. Global or continuous optimisation has attracted some of the above such as GAs but also others like HS, DE, and particle swarm. The latter category is also commonly and successfully applied in engineering areas, whereas the former is more known for solving managerial and business problems. To complete the chapter, some of the other population methods including recent ones are also briefly highlighted. The reader may found one of these techniques, though relatively less popular or newer, to be more exciting and with more scope for future investigation. The next chapter will deal with hybridisation.

References

Alp, O., Erkut, E., & Drezner, Z. (2003). An efficient genetic algorithm for the p-median problem. *Annals of Operations Research, 122*, 21–42.

Bakhouya, M., & Gaber, J. (2007). An Immune inspired-based optimisation algorithm: Application to the travelling salesman problem. *Advanced Modeling and Optimization, 9*, 105–116.

Bullnheimer, B., Harlt, R., & Strauss, C. (1998). Applying ant systems to the vehicle routing problem. In S. Voss, S. Martello, I. H. Osman, & C. Roucairol (Eds.), *Metaheuristics: Advances and trends in local search paradigms for optimization* (pp. 285–296). Boston: Kluwer.

Caserta, M., & Quinonez Rico, E. (2009). A cross entropy-based metaheuristic algorithm for large scale capacitated facility location problems. *The Journal of the Operational Research Society, 60*, 1439–1448.

de Castro, L. N., & Timmis, J. (2002). *Artificial immune systems: A New Computational Intelligent Approach*. London:Springer.

Colorni, A., Dorigo, M., & Maniezzo, V. (1991). Distributed optimization by Ant Colonies. In F. Varela & P. Bourgine (Eds.), *Proceedings of the European conference on artificial life* (pp. 457–474). Amsterdam: Elsevier Publishing.

Daneubourg, J. L., Aron, A., Goss, S., & Pasteels, J. M. (1990). The self organising exploratory pattern of the argentine ant. *Journal of Insect Behavior, 3*, 159–168.

Das, S., Abraham, A., Chakraborty, U. K., & Konar, A. (2009). Differential evolution with a neighbourhood-based mutation operator. *IEEE Transactions on Evolutionary Computation, 13*, 526–553.

Dasgupta, D. (Ed.). (1999). *Artificial immune system and their applications*. Berlin/New York: Springer.

De Jong, K. A. (1975). *An analysis of the behavior of a class of genetic adaptive systems*. Doctoral dissertation, University of Michigan, Ann Arbor.

Dorigo, M., & Gambardella, L. M. (1997). Ant colony system: A cooperative learning approach to the travelling salesman problem. *IEEE Transactions on Evolutionary Computation, 1*, 53–66.

Dorigi, M., & Stutzle, T. (2010). Ant colony optimization: Overview and recent advances. In M. Gendreau & J. Y. Potvin (Eds.), *Handbook of metaheuristics* (2nd ed., pp. 227–264). London: Springer.

Dorigo, M., Caro, G., & Gambardella, L. (1999). Ant algorithms for discrete optimization. *Art Life, 5*, 137–172.

Fletcher, R. (1987). *Practical methods of optimisation* (2nd ed.). Chichester: Wiley, 2001 reprint.

Garcia-Villoria, A., & Pastor, R. (2009). Introducing dynamic diversity into a discrete particle swarm optimization. *Computers and Operations Research, 36*, 951–966.

Geem, Z. W., Kim, J. H., & Loganathan, G. V. (2001). A new heuristic optimization algorithm: Harmony search. *Simulation, 76*(2), 60–68.

Glover, F., & Laguna, M. (1997). *Tabu search*. Boston: Kluwer.

Glover, F., Laguna, M., & Marti, R. (2003). Scatter search and path relinking: Advances and applications. In F. Glover & G. A. Kochenberger (Eds.), *Handbook of metaheuristics* (pp. 1–35). London: Kluwer.

Goldberg, D. E. (1989). *Genetic algorithm in search, optimization and machine learning*. New York: Addison-Wesley.

Goldberg, D. E., & Lingle, R. (1975). Alleles, loci and the travelling salesman problem. In J. J. Grefenstette (Ed.), *Proceedings of an international conference on genetic algorithms and their applications* (pp. 154–159). Hillsdale: Lawrence Erlbaum Associates.

Gomez, A., Amran, I., & Salhi, S. (2013). Solution of classical transport problems with bee algorithms. *International Journal of Logistics Systems and Management, 15*, 160–170.

Holland, J. H. (1975). *Adaptation in natural and artificial systems*. Ann Harbor: University of Michigan Press.

Ishida, Y., Hirayama, H., Fujita, H., Ishiguro, A., & Mori, K. (Eds.) (1998). *Immunity-based systems-intelligent systems by artificial immune systems*. Tokyo: Corona Pub. Co.

Karaboga, D. (2005). *An idea based on honey bee swarm for numerical optimization*. Technical Report TR06, Computer Engineering Department, Erciyes University.

Kennedy, J., & Eberhault, R. C. (1995). Particle swarm optimization. *IEEE international conference on neural networks*, Perth, pp. 1942–1948.

Lu, H., & Chen, W. (2006). Dynamic-objective particle swarm optimization for constrained optimization problems. *Journal of Combinatorial Optimization, 12*, 408–419.

Mahdavi, M., Fesanghary, M., & Damangir, E. (2007). An improved harmony search algorithm for solving optimization problems. *Applied Mathematics and Computation, 188*, 1567–1579.

Marti, R., Laguna, M., & Glover, F. (2006). Principles of scatter search. *European Journal of Operational Research, 169*, 359–372.

Maslow, A. H. (1954). *Motivation and personality.* New York: Harper & Sons.

Neri, F., & Tirronen, V. (2010). Recent advances in differential evolution: A survey and experimental analysis. *Artificial Intelligence Review, 33,* 61–106.

Pan, Q. K., Suganthan, P. N., Tasgetiren, M. F., & Liang, J. J. (2010). A self-adaptive global best harmony search algorithm for continuous optimization problems. *Applied Mathematics and Computation, 216,* 830–848.

Pham, D. T., Ghanbarzadeh, A., Koc, E., Otri, S., Rahim, S., & Zaidi, M. (2006). The bees algorithm, a novel rool for complex optimisation problems. In *Proceedings of the 2nd virtual international conference on intelligent production machines and systems,* Elsevier, pp. 454–459.

Resende, M. G. C., & Ribiero, C. G. (2005). Scatter search and path-relinking: Fundamentals, advances, and applications. In M. Gendreau & J. Y. Potvin (Eds.), *Handbook of metaheuristics* (2nd ed., pp. 87–107). London: Springer.

Rubinstein, R. Y. (1997). Optimization of computer simulation models with rare events. *European Journal of Operational Research, 99,* 89–112.

Rubinstein, R. Y., & Kroese, D. P. (2004). *The cross-entropy method: A unified approach to combinatorial optimization, Monte Carlo simulation and machine learning.* New York: Springer.

Salhi, A., & Fraga, E. S. (2011). Nature-inspired optimisation approaches and the new plant propagation algorithm. In *Proceedings of the ICeMATH2011,* pp. K2–1 to K2–8.

Salhi, S., & Gamal, M. D. H. (2003). A genetic algorithm based approach for the uncapacitated continuous location-allocation problem. *Annals of Operations Research, 123,* 203–222.

Salhi, S., & Petch, R. (2007). A GA based heuristic for the vehicle routing problem with multiple trips. *Journal of Mathematical Modelling and Algorithms, 6,* 591–613.

Selamoglu, B. I., & Salhi, A. (2016). The plant propagation algorithm for discrete optimisation: The case of the travelling salesman problem. In X. S. Yang (Ed.), *Nature-inspired computation in engineering* (Studies in computational intelligence, Vol. 637, pp. 43–61). Switzerland: Springer.

Storn, R., & Price, K. (1997). Differential evolution- a simple and efficient adaptive scheme for global optimization over continuous spaces. *Journal of Global Optimization, 11,* 341–359.

Stützle, T., & Hoos, H. H. (2000). MAX-MIN ant system. *Future Generation Computer Systems, 16,* 889–914.

Sulaiman, M., Salhi, A., & Selamoglu, B. I. (2014). A plant propagation algorithm for constrained engineering optimisation problems. *Mathematical Problems in Engineering, 2014*, 1–17.

Sulaiman, M., & Salhi, A. (2016). A hybridisation of runner-based and seed-based plant propagation algorithms. In X. S. Yang (Ed.), *Nature-inspired computation in engineering* (Studies in computational intelligence, Vol. 637, pp. 1–18). Switzerland: Springer.

Szeto, W. Y., Wu, Y., & Ho, S. C. (2011). An artificial bee colony algorithm for the capacitated vehicle routing problem. *European Journal of Operational Research, 215*, 126–135.

Tarantilis, C. D., & Kiranoudis, C. T. (2002). BoneRoute: An adaptive memory-based method for effective fleet management. *Annals of Operations Research, 115*, 227–241.

Tiwari, M. K., Prakash, A., Kumar, A., & Mileham, A. R. (2005). Determination of an optimal sequence using the psychoclonal algorithm. *Journal of Engineering Manufacture, 219*, 137–149.

Wade, A. C., & Salhi, S. (2003). An ant system algorithm for the mixed vehicle routing problem with backhauls. In M. G. Resende & J. P. de Sousa (Eds.), *Metaheuristics: Computer decision-making* (pp. 699–719). New York: Kluwer.

Wedde, H. F., Farooq, M., & Zhang, Y. (2004). BeeHive: An efficient fault-tolerant routing algorithm inspired by honey bee behavior. In *Ant colony optimization and swarm intelligence* (Lecture notes in computer science, Vol. 3172, pp. 83–94). Berlin: Springer.

Yang, X.-S. (2005). Engineering optimizations via nature-inspired virtual bee algorithms. In J. Mira & J. R. Alvarez (Eds.), *IWINAC 2005* (Lecture notes in computer science, Vol. 3562, pp. 317–323). Berlin/Heidelberg: Springer.

5

Hybridisation Search

5.1 Hybridisation of Heuristics with Heuristics

The usual integration of heuristics is usually a two-phase method where one heuristic calls the other, at the end making the latter behaving as a simple post-optimiser. This approach is now extended to incorporate the integration within each of the heuristic/metaheuristic. This view seems to attract more attention in recent years. The most used ones include hyper-heuristics and memetic algorithm (MA).

Hyper-Heuristics

This type of heuristics hybridisation is originally designed to construct a feasible solution of a good quality using several instead of one constructive/greedy heuristic only. The reasoning behind it is that each heuristic has <u>strengths and weaknesses,</u> and it is thought to be appropriate if those aspects are taken into account during the search. Though this notion has been hinted in the literature from time to time since the 1960s (Ross 2005), these methods are intensively studied by Burke et al. (2003). There

© The Author(s) 2017   **129**
S. Salhi, *Heuristic Search*, DOI 10.1007/978-3-319-49355-8_5

are two ways in implementing this kind of approach which I refer to as constructive and improvement hyper-heuristics.

Constructive Hyper-Heuristics

The idea is to use, at each iteration during the construction of the solution, a chosen heuristic from a list of low-level heuristics either randomly or intelligently. Low-level heuristics are usually simple greedy heuristics that are commonly used to produce reasonably good quality solution rather quickly. These are usually then used as a starting point in other powerful heuristics such as those described in the previous chapters. The design of such a high level heuristic that controls the evolution of the incomplete solution and hence decides which low-level heuristics is more appropriate at that very iteration is known as hyper-heuristic. The following schemes are worth discussing:

A Brute Force Approach

This simplest approach is to apply all the low-level heuristics at each step of the construction of the solution, and choose the one that produces the best solution by selecting the next attribute from that best configuration to add to the incomplete solution configuration. The process is repeated until all attributes are included to make up a complete feasible solution. This mechanism guarantees that the obtained solution is either better or the same as the solution found by the best individual low-level heuristic.

A Random Scheme

The selection is to choose at each step of the construction of the solution configuration, one of these low-level heuristics randomly and with equal probability $\frac{1}{p}$ with p denoting the number of low-level heuristics. This is obviously much quicker than the Brute Force Method but may produce a solution which is inferior to the one that could be generated by the best low-level heuristic.

A Learning-Based Approach

This is a two-stage process. In stage one (i.e., the training stage), each low-level heuristic is applied over the set of instances (or just over a training subset) and the frequency of producing the best solution or the probability based on the deviations from the best solutions is recorded. A minimal threshold can be imposed if the best solutions were found by a tiny number of these heuristics only. In stage two, this information is then used to choose the low-level heuristic at each step of the construction of the new solution as is used in the above two schemes. This selection is performed pseudo-randomly based on the performance of these low-level heuristics at the first stage. The only weakness is that the approach needs to run more than once or at least on a smaller subset of instances which lead to biases towards the selected subset. Another slightly more fine-tuned version is based on the brute force approach to record the information in stage one which can be performed based on the first steps that are used to construct the incomplete solutions only. This information is then passed to stage two where in subsequent steps the selection of the low-level heuristics is performed pseudo-randomly. This approach is quicker than the brute force method but could produce relatively less competitive results as it may miss some good opportunities when selecting the next attribute.

Note that both the first and the third methods could be made more focussed if a smaller subset of promising low-level heuristics that showed to produce good but diverse solutions are chosen a priory so to avoid those heuristics that are proved to behave poorly in most cases.

Improvement Metaheuristics

Here, an initial solution is already found and the use of hyper-heuristic is adopted to act as a post-optimiser or a local search. However, instead of using one particular metaheuristic, say TS or SA, to improve the solution, the hyper-heuristic philosophy is adopted. Note that it is common and simpler to use local searches as low-level heuristics in these type of hyper-heuristics though the use of metaheuristics is also attempted by a few authors. One way would be to split the time horizon into cycles where in each cycle we can have a learning phase and a launching phase, each split into several iterations or short periods. In the learning phase, all the metaheuristics are used in those short periods and the frequencies of obtaining improved solutions are recorded. This information is then used in the launching stage to select pseudo-randomly at each short period one metaheuristic. Here, the question is how to spread the total computing time between each run of a given metaheuristic and how long the short periods in the training and the launching stages should be. For instance, Garcia-Villoria et al. (2011) investigated both approaches and obtained superior results with the latter when solving a class of scheduling problems.

This approach is displayed in Fig. 5.1.

It is worth noting that the use of hyper-heuristic could also be considered as a form of an expert system that guides iteratively the selection of the heuristics within the search in order to solve a given optimisation problem, irrespective whether these heuristics are low or high level. A similar idea based on the second approach and which is commonly used in most metaheuristics is the choice of the best move from a few different neighbourhoods such as the add, the drop and the swap moves. The move that produced the best solution is then implemented at that given iteration and the process continues until some stopping criteria are satisfied. This is similar to the crude approach but using improvement-based schemes instead. This naïve approach may not be useful when using metaheuristics as the whole process will be taking too long. The strategy for selecting the low-level heuristic at a given iteration and the choice of the pool of these low-level heuristics to be used can be challenging especially if this kind of heuristics is considered to operate at a higher level, and to be quick and independent of the problem studied. Burke et al. (2010) give an interesting classification of hyper-heuristics and provide further references on this issue, with a focus on time tabling problems.

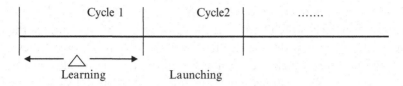

Fig. 5.1 A representation of the improvement hyper-heuristics

MA and Its Variants

These are based on a population of solutions like GA except that each solution (chromosome) within the same generation is improved via intensification by a refinement/improvement procedure like a local search. MAs can therefore be considered more aggressive than their classical GAs. Obviously this combination can be seen as a hybridisation of GA and local search. Such a heuristic was originally put forward by Moscato and Norman (1992) and an overview of this approach can be found in Moscato (1999) and Moscato and Cotta (2003) though other authors have also implemented a similar strategy indirectly, see, for instance, Thangiah and Salhi (2001) for the case of multi-depot routing. Here, the multi-level heuristic of Salhi and Sari (1997) was integrated with GA to improve the obtained GA chromosomes at each generation. The questions may include how long the improvement procedure is allowed to run? Is local search necessary at each generation or just for specific chromosomes? And how to avoid the risk of early stagnation due to such intensifications carried out by local search? To control diversity and to answer the last question, Sorensen and Sevaux (2006) introduced schemes to manage the population (which they refer to as MA/PM). An efficient implementation of MA/PM for solving an integrated distribution/production system is also proposed by Boudia and Prins (2009). The above authors put emphasis on how to better manage the population so to maintain diversity and to guide the search more efficiently. The idea is to use a smaller population size than in GA (say 20 or 30), and also to control diversity, not through mutation as in GA, but via a measure of diversity which is embedded into the generation of the chromosomes, such as dissimilarity or distance between solutions. Initially most chromosomes are improved via a local search (usually with a high probability).

Here a crossover operator is used only and the obtained child is then improved by local search with a smaller probability. Such a child is then accepted only if it is diverse enough from the current chromosomes or happens to yield a much better value than the current best. This measure of dissimilarity is also similar to the one used in SS as previously described in Chap. 4. In simple terms, MA/PM can be considered as a guided MA, though it has more parameters than GA and the classical MA. One may argue that this avenue may not be worth pursuing due to its additional parameters and computational burden, but, in my view, this has scope for improvement and hence could lead to much more promising outcome. For the context of multi-objective, which will be discussed in the next chapter, Elhossini et al. (2010) develop an hybridisation approach of two evolutionary methods for determining an efficient approximation of the Pareto frontier in multi-objective non-linear optimisation. They integrate PSO with an evolutionary method such as GA where these are activated in a simple post optimisation manner (GA after PSO and vice versa).

A Brief on Other Metaheuristic Hybridisations

The above population-based hybridisation is also adopted in other population-based heuristics described in Chap. 4 as these are successful at finding promising regions where good solutions may lie but not powerful enough to localise the exact local optima. The latter task is achieved using some form of local searches or those one point solution methods though some population-based like PSO have also shown to be worth exploring for this particular task. For instance, a local search or small runs of those metaheuristics described in chapters two and three could be combined with ACO after global updating is performed, with the bee algorithm ABC once some sites are dropped or reinforced, with PSO when global velocity update is activated, with HS once the new set of harmonies are selected and so on. The interested reader could consider any of the above by thoroughly investigating the time to activate the local search, how intensive this is implemented and which solutions are better suited for such improvement. The choice of the subset of solutions, if not all are selected, can be based on the top quality solutions as well as those with high level diversity.

Other integrations of metaheuristics have also been attempted. For instance, for the vehicle routing problem, interesting results were obtained by Xiao et al. (2014) who integrate VNS with SA, and by Akpinar (2014) who consider the hybridisation of LNS with ACO. For continuous optimisation, hybridisation of evolutionary methods with others is also commonly adopted. For instance, Norman and Iba (2008) enhanced their DE heuristic by incorporating a local search based on a hill climbing Simplex Crossover on certain solutions at a given generation. Another interesting integration of DE this time with another evolutionary approach instead namely PSO was successfully applied for tackling multi-objective continuous optimisation problems by Mashwani and Salhi (2010). Here, at each generation, part of the population is tackled by PSO, whereas the remaining part by DE with the partition being dynamically adjusted. For more information on hybridisation of metaheuristics, the review by Raidl (2006) can be a good reference.

5.2   Integrating Heuristics Within Exact Methods

Exact methods can be slow and inappropriate for some applications but could be made useful if combined with heuristic search. In this section I shall present of some schemes that are shown to be promising and worth considering.

Injection of UB

The simplest way is to introduce the value of a feasible solution from a heuristic as an UB say within B&B. This extra information will fathom any unnecessary branches from being explored. Note that this additional single constraint may incur an extra computational effort.

Tightening of LB and UB

There are some approaches that aim to tighten both the UBs and the LBs simultaneously. Some of these will be discussed here.

Lagrangean Relaxation Heuristics (LRH)

This approach takes into account the advantages of both exact and heuristic methods for generating proven LBs and corresponding UBs, respectively. This type of hybridisation could be discussed as part of the next section but due to its special structure, it is simpler to consider it on its own right here. This type of heuristics are also classified under mathematically based heuristics, see Salhi (2006).

LRH was first proposed by Held and Karp (1970, 1971) and proved to be successful at solving many classes of optimisation problems. The main concept of LRH is to identify the set of constraints of a general integer program that resulting in an increase of the computational complexity of the problem making it intractable. These constraints are then introduced into the objective function in a Lagrangean fashion by attaching to them unit penalties, that will be reduced as the search progresses with the aim to reducing the amount of feasibility violation. This transformation should be constructed to render the new relaxed problem easier to solve optimally (usually by inspection or basic polynomial optimal type algorithms) and hence produce LBs for the original problem. Note that the relaxed problem could, after mathematical manipulations, turn into a subset of separable subproblems, instead of a single one only. These subproblems are solved individually and their respective optimal objective function values and their corresponding decision variables values are then used to make up the overall solution for the relaxed problem. A feasible solution of the original problem (an UB in case of a minimisation problem) is then derived using a usually quick heuristic method. The penalties are adjusted based on the violation and the process is repeated until one of the stopping criteria is met. This could include, for instance, the gap between the best lower and UB is small enough, a negligible change in the solution configuration is detected, the step size is nearly zero, the maximum

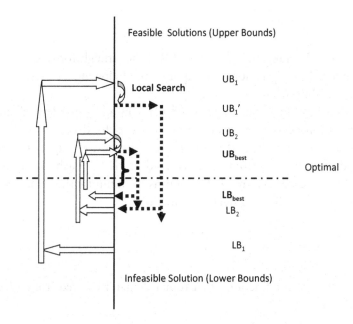

Fig. 5.2 Principle of Lagrangean relaxation heuristics

computing time is reached, among others. A schematic graph of LRH is given Fig. 5.2.

Theoretically after several iterations this approach should lead to a near or optimal solution with a duality gap showing the quality of the solution. In other words, we have

$\lim_{k \to \infty} |UB(k) - LB(k)| \to 0$ if the penalties are ideally found.

Note that in Fig. 5.2, an additional mini step that aims to improve the UB that is transformed from the LB is also introduced. A simple local search could be used for this purpose. This yields a tighter UB which in turn will speed up the process when adjusting the penalties and so on. This mechanism was successfully applied for a class of discrete location problems by Agar and Salhi (1998) which is described next for completeness.

Illustrative Example

As the use of Lagrangian Relaxation (LR) is not straightforward and some mathematical manipulations are required, the following location problem is used for illustration.

Consider the following capacitated plant location problem where we have m potential sites, n customers to serve, a fixed cost of opening a facility at site i, say F_i, and the transportation cost of supplying all the demand of customer j from a facility sited at i is C_{ij}. Each facility is also limited by its capacity, say Q_i. Our aim is to minimise the total cost.

Consider the following decision variables:

$$Y_i = \begin{cases} 1 & \text{if a facility is located at site } i \\ 0 & \text{otherwise} \end{cases}$$

and X_{ij} denotes the fraction of customer demand j served from a facility located at i.

$$(P) \begin{cases} \text{Min} \sum_{j=1}^{n}\sum_{i=1}^{m} C_{ij}X_{ij} + \sum_{i=1}^{m} F_i Y_i & (5.0) \\ st \\ \sum_{i=1}^{m} X_{ij} = 1\, j = 1, \ldots, n & (5.1) \\ \sum_{j=1}^{n} q_j X_{ij} \leq Q_i Y_i\, i = 1, \ldots, m & (5.2) \\ Y_i \in \{0, 1\} \text{ and } 0 \leq X_{ij} \leq 1\, i = 1, \ldots, m \text{ and } j = 1, \ldots, n & (5.3) \end{cases}$$

The objective function is represented by (5.0) defining the total cost that includes the fixed cost and the variable cost. Constraints (5.1) guarantee that every customer is fully served, constraints (5.2) show that the capacity constraint at an open facility is not exceeded while guaranteeing that a customer is served from an open facility only. The last constraint (5.3) is the binary and the non-negativity constraints. If (5.1) and (5.2) are relaxed and transferred to the objective function the remaining problem may be easy to solve. Let s_i and t_j be the Lagrangean multipliers (penalties) associated with (5.1) and (5.2), respectively. The new objective function becomes

$$\text{Min} \sum_{j=1}^{n} \sum_{i=1}^{m} C_{ij}X_{ij} + \sum_{i=1}^{m} F_i Y_i + \sum_{j=1}^{n} s_j \left(1 - \sum_{i=1}^{m} X_{ij}\right) + \sum_{i=1}^{m} t_i \left(-Y_i + \frac{\sum_{j=1}^{n} q_j X_{ij}}{Q_i}\right)$$

$$= \sum_{i=1}^{m} \sum_{j=1}^{n} \left[C_{ij} + \frac{T_i}{Q_i}q_j - s_j\right]X_{ij} + \sum_{i=1}^{m} (F_i - t_i)Y_i + \sum_{j=1}^{n} s_j$$

Note that if a facility at site i is opened, its contribution would be

$$a_i = F_i - t_i + \sum_{j=1}^{n} \text{Min}\left(0, \alpha_{ij}\right) \text{ where } \alpha_{ij} = C_{ij} + \frac{t_i}{Q_i}q_j - s_j.$$

Therefore, the LB problem reduces to solving the following:

$$(LB) \begin{cases} \text{Min} \sum_{i=1}^{m} a_i Y_i + \sum_{j=1}^{n} s_j \\ st\, Y_i \in \{0, 1\} \end{cases}$$

Given some values of (s_j) and (t_i), LB can be solved optimally by inspection by setting $Y_i = 1$ when $a_i < 0$ leading to the obtained LB value $Z(LB)$. In other words, let $\{i = 1, \ldots, m$ such that $Y_i = 1\}$ be the set of open facilities. Consider these facilities as the p open facilities (i.e., $p = \left|\{(i = 1, \ldots, m) \text{ such that } Y_i = 1\}\right|$), and for simplicity of notation, reorder the indices such that the first p are the open facilities. The problem then reduces to a standard transportation problem (UB).

$$(UB) \begin{cases} \text{Min} \sum_{i=1}^{p} \sum_{j=1}^{n} C_{ij}X_{ij} \\ \text{such that} \\ \sum_{j=1}^{n} q_j X_{ij} \leq Q_i \ \ i = 1, \ldots.p \\ 0 \leq X_{ij} \leq 1\, j = 1, \ldots, n \text{ and } i = 1, \ldots, p \end{cases}$$

for which an optimal solution (heuristic solution if necessary) of the X_{ij} can be easily obtained which leads to an UB $Z(UB)$. Note that this could be improved by a local search in opening/closing some facilities from the set of open facilities and solving the *TP* again. The violations in the constraints (5.1) and (5.2) defined by $G_j = 1 - \sum_{i=1}^{m} X_{ij}$ and $H_i = -Y_i$

$+ \dfrac{\sum_{j=1}^{n} q_j X_{ij}}{Q_i}$ respectively are then computed. The Lagrangean multipliers are updated as follows

$$s_j = s_j + T.G_j \text{ and } t_i = \text{Max}(0, t_i + T.H_i) \text{ where } T$$
$$= \frac{\lambda(\beta Z(UB) - Z(LB))}{\gamma(\sum_{j=1}^{n} G_j^2 + \sum_{i=1}^{m} H_i^2)} \ .$$

With $\lambda = 2$ initially and halved after a certain number of iterations and $\beta = 1.05$ just over 1 to provide flexibility to the UB value in case the gap quickly gets smaller and hence less informative as the step size T may go to zero too quickly. In addition, the correction factor which is usually set to 1 is replaced here by $\gamma = 8$ which is found to be very efficient in reducing the total number of iterations and tightening the gap (see Agar and Salhi 1998). The process is then repeated until certain criteria are met.

It is worth noting that not all problems fit nicely into this LRH formulation as both the LB problem and its corresponding feasible problem (UB) may not be always easy to solve. It is also worth mentioning that there is no guarantee that the new corresponding UB found at a given iteration of LRH is always an improving feasible solution. This is due to the heuristic procedure used to transform the LB into a feasible solution. In this case extra, care is required to guide the search by referring to the best UB so far when updating the Lagrangean multipliers and so on. For further background, the reader will find the seminal paper by Geoffrion (1974) and the studies by Beasley (1993) and Agar and Salhi (1998) to be interesting and informative.

Integration of LRH Within Heuristics/Metaheuristics As part of an exact method in the sense that both the LBs and UBs can be generated and hence duality gap can be measured, LRH is incorporated within heuristics to solve a subproblem within the larger problem in a recursive manner. For instance, very recently Bouzid et al. (2016) embedded LRH to split a giant tour (the TSP) into smaller vehicle tours in an optimal or near optimal manner, where VNS is then used as a diversification process to get a new giant tour. This implementation showed to be much more efficient than using the classical ILP formulation to optimally split the giant tour. Note that there are obviously other polynomial type optimal algorithms to split a giant tour for the case of the VRP such as DP and the well-known network optimisation algorithm—the Dijkstra's algorithm. However it is worth stating that these fast methods are restricted to such type of problems only and are not easily extended to cater for other variants of splits such as the presence of time windows, fixed number of vehicles and so on.

Reduction/Relaxation-Based Schemes

The idea is based on a dynamic variable fixation mechanism obtained by solving a series of relaxed problems with an efficient way of recording useful information using memory and frequency. For instance, a solution that is never used is more likely not to be used and a variable which has used most often could be fixed permanently. This is an iterative process that continues until no change in the solution is found.

One approach, which seems to have gained more pace recently, is to solve a series of reduced problems exactly by fixing some of the variables to their integer values as found by the *LP relaxation*. The problem is then augmented by an additional constraint every time leading to a new LB and hence to a new fixation of the variables for the new reduced problem. This process is then repeated until the gap between LB and UB is either zero or negligible. This concept is adopted by Hanafi and Wilbaut (2006) for solving the multi-dimensional knapsack problem. Wilbaut et al. (2006) also incorporated DP with TS where DP is used on reduced problems and TS is applied to the remaining variables. An interesting review paper by

Wilbaut et al. (2008) discusses some of these aspects with a focus to knapsack problems.

A similar view that is based on solving a succession of subproblems optimally while adding at each iteration a few attributes (say r) is successively presented by Chen and Chen (2009) for solving both the continuous and the discrete p-centre problems. This approach though sensitive to the values of r has the advantage that the optimal solution if it exists can be identified whenever the optimal solution of the last reduced problem happens to be feasible for the original one. One approach that also incorporates solutions and useful information from heuristics to tighten both the UB and LBs in the corresponding ILP formulation is presented by Salhi and Al-Khedhairi (2010) for the vertex p-centre problem.

Kernel Search

This approach is a relatively recent iterative approach introduced by Angelelli et al. (2010) for solving MILP problems. The idea is to define a succession of smaller related MILP problems where promising values, usually those with positive values, are used to make up a reduced and smaller problem with the idea that at each iteration, additional variables are added to the subproblem, whereas some may be removed. At the initial phase, the LP is usually solved and all non-negative variables are used to make up the subset of promising variables which is referred to as the kernel from which Kernel Search (KS) was named. The rest of the variables are put into groups known as buckets. At each iteration one bucket is added to make up a bigger problem. This approach is adapted to allow for the removal of some variables from the kernel and the sequence in which the buckets are added, For instance, the variables to be removed from the kernel can be those which happen not to be part of the optimal solutions in the last few iterations. This removal strategy can be extended to identify a performance measure for the variables based on frequency of occurrence, their shadow prices if not selected and so on. This process continues until the last bucket is utilised. Guastaroba and Speranza (2012) explore these issues for the case of large capacitated facility location problem with competitive results. In this iterative approach, at each iteration, the new objective function value, if it is found to be an improved and feasible solution, is considered as the new UB which is then added to tighten the formulation. The choice of the bucket to add and the selection of the variables to remove from the kernel constitute a challenging part of this approach that could be critical to the success of KS.

Regular Seeking of New UBs

Another option which is more challenging is to integrate heuristic and exact methods more intensively by calling a heuristic which (a) transforms an infeasible solution (LB) into a feasible one at <u>certain nodes of the B&B tree</u> and (b) improves the best UB or even other promising but less-attractive already found UBs. The question is to determine which nodes are worthwhile exploring and also which type of heuristics are better suited for (a) or (b).

A basic representation of (a) and (b) is shown in Fig. 5.3.

In other words, instead of applying the optimal method (B&B, Branch & Cut) till the end, we can inject extra information at intermediate points of the search that need to be identified. For instance, in B&B, one way would be to take a solution at a given node and see whether that solution (UB or LB) can be either improved or transformed from LB to UB. For instance, if the node is a feasible node as mentioned in (b) with a value UB_0 (see Fig. 5.3), a metaheuristic or just a simple heuristic that uses that feasible solution as a starting point may improve it (say UB_0^+), and hence provide a tighter UB for quick fathoming. This mechanism can be either applied each time a new UB within B&B is obtained even if it is not better than the best, or if the new UB happens to be slightly worse than the best incumbent UB but passes a certain threshold, as commonly used in TA (see Chap. 3). The new improved UB, if it exists, may be proved to be even optimal if it happens to match the best LB so far or satisfies certain optimality rules. Similarly in (a), if an infeasible solution is at a node (LB_1), before branching a repair mechanism that transforms such an infeasible solution (i.e., LB) into a feasible one (say UB_1^0) could be used. If this solution happens to be better than the best UB so far, this will

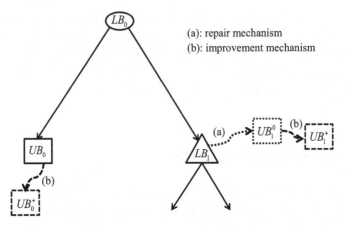

Fig. 5.3 Integrating heuristic power within B&B

become the incumbent UB leading to further fathoming in the B&B tree. This temporary node with UB_1^0 can then be improved via (b) again to obtain an even tighter bound, say UB_1^+. One obvious way would be to choose those attributes with higher values close to 1 (in case of binary optimisation) while checking feasibility.

Local Branching

There are however interesting studies that aim to find feasible solutions, the first of which is the idea of local branching (LB) developed by Fischetti and Lodi (2003). For simplicity, consider a binary optimisation problem where a LB is found at a given node (including the source node) with $\widehat{X}_i = [X_i^*]$ for all $i \in E_1$ where the set $E_1 \subset E$ is the required subset of binary variables and $\widehat{X}_i = X_i^*$ for all $i \in E - E_1$ (those not required to be binary or integer say). The idea is to solve the original problem with the addition of the following linear constraint given a suitable value of k which aims to restrict the distance between X and $\widehat{X}$ to be within k only using $\Delta\left(X,\widehat{X}\right) = \sum_{i \in E_1 : \widehat{X}_i = 0} X_i + \sum_{i \in E_1 : \widehat{X}_i = 1} X_i \leq k$. This is updated and the process is repeated until a feasible solution is found. In other words, if the solution is not improved after a certain time limit at a node, k is halved, however if no solution was found in that time limit or the solver proved infeasibility, the new linear constraint is relaxed by setting $k = k + \frac{k}{2}$. An enhanced method, which avoids some of these parameters, is known as feasibility pump (FP). This is presented by Fischetti et al. (2005) whose idea is to solve the auxiliary subproblem by substituting the original objective function with $\text{Min} \, \Delta\left(X, \widehat{X}\right) = \sum_{i \in E_1 : \widehat{X}_i = 0} X_i + \sum_{i \in E_1 : \widehat{X}_i = 1} X_i$ instead. This is repeated until a feasible solution is obtained. Extra refinements to reduce the computing time and to get a good quality solution are also examined. As both LB and FP run under a certain time limit, the obtained solution if it exists is obviously a feasible MIP solution. Very competitive results are produced for a variety of difficult ILP problems, see Fischetti et al. (2005) for more detail. An interesting adaptation of LB by Hansen et al. (2006) is to use a general VNS structure within LB. Here, the initial step at the root node is found as in LB as this is the initial solution found by a commercial solver such as CPLEX. In the shaking step of the VNS, the values of the parameter of k follow the VNS structure instead where the kth neighbourhood is defined by:

$$N_k\left(\widehat{X}\right) = \left\{X \mid \Delta\left(X, \widehat{X}\right) \leq k\right\} ; k = K_{\min}, \ldots., K_{\max}.$$

In the local search, a VND is applied where a solver is also used to find the local minimum based on $N_l\left(\widehat{X}\right); l = 1, \ldots., l_{\max}$ defined as the new linear constraint. Very competitive results are discovered when compared against the classical LB results.

Other examples that use such a similar hybridisation include the work by Nwana et al. (2005) who combined B&B with SA to solve the binary and the mixed ILPs. A depth first of B&B is incorporated to generate an initial feasible solution and SA is used to determine feasible ones whenever infeasibility is encountered. This is a hybrid approach that is based on sharing information between concurrent runs of B&B and SA. Wilbaut et al. (2009) combine heuristic solutions and ILP formulations to design an adaptive search for solving the 0–1 knapsack problem. Lazic et al.

(2010) also embedded VNS with ILP formulation to tackle 0–1 mixed integer programs. For more information on the integration of metaheuristics with mathematical programming, see the interesting review by Talbi (2016) who also discusses constraint programming as well as machine learning.

5.3   Integration of ILP Within Heuristics/ Metaheuristics

In some cases, the original problem reduces to manageable and easy subproblems to be solved optimally when the promising attributes are identified only or when the problem is split into separable subproblems that are also easy to solve optimally. I present one approach for the former (heuristic concentration [HC]) and a brief on the latter (problem decomposition).

Problem Decomposition

In some cases, the original problem can be naturally decomposed into separable subproblems, usually two, which are solved optimally once one is fixed. As the output of one subproblem becomes the input of the other, the original problem cannot be guaranteed to be optimally solved. To enhance the chance of getting a good solution, a recursive approach is usually adopted. As an example solving the capacitated or the single source location problem, once the facilities are sited the problem reduces to either a transportation problem (TP) or an generalised assignment problem (GAP). In the former, a customer may not need to be served from one facility only, whereas in the latter, a customer is entirely served by one facility only. Both the TP and the GAP can be solved using standard commercial optimisation solvers such as CPLEX, LINDO, Xpress-MP and GuRobi just to cite a few. The metaheuristic is used to relocate some of the facilities. As the number of calls to the TP (or the GAP) can be too excessive, an approximation technique or heuristic could be used instead and once in a while the allocation is performed optimally. The balance between the use of approximation and optimality depends on the complexity of the ILP problem and the problem size. As an example the TP is relatively much easier to solve than the GAP so the number of calls to optimally solve the TP could be used more often than its counterpart the GAP if found necessary.

A similar example, commonly encountered in routing strategies, is the cluster-first routing-second strategy. Here, the clusters are first constructed and then in each cluster a small TSP is optimally solved. The search is to guide the search by identifying other sets of clusters usually through performing a certain level of perturbations on existing clusters.

Another approach that also deals with the same routing problem is to remove from a feasible tour some sequences, as in ILS, and then insert them back optimally to the split points of the tours through solving optimally the corresponding set partitioning problem where this ruin/repair mechanism is repeated several times. This approach is successfully adopted by De Franceschi et al. (2006).

Heuristic Concentration (HC)

This is a two-stage process where in Stage 1, a large number of solutions are generated randomly using multi-start, for instance, or through a heuristic such as GRASP. The selected attributes of those solutions are then put together to make up the concentration set which is obviously a relatively much smaller subset. The aim of the selected subset is to represent the original problem with those promising attributes only without considering the large number of irrelevant ones. The reasoning behind this idea is that the optimal or the final best solution is likely to consist of those attributes. Note that HC shares some similarities with CE in terms of identifying the subset of promising attributes. In Stage 2, an exact method or an intensive-based metaheuristic is then applied on this small subset. This idea is originally developed by Rosing and ReVelle (1997) with some variants being later proposed by Rosing et al. (1999) for solving the p-median problem. For instance, in Stage 1, only those attributes (potential sites) that belong to all the solutions are considered as fixed and hence will be part of the optimal solution and in Stage 2, a reduced problem, made up of these potential selected sites, is then solved either optimally or heuristically with a powerful metaheuristic. Another stricter way is to choose the concentration set from those attributes that happen to exist in $k(k > 1)$ times in the solutions configurations. The reasoning behind this is that some attributes with very low frequency of occurrence (i.e., $k = 1$) could have been chosen by luck and hence may distort the basic principle of this method. The value of k can affect the final solution and hence linking k to the number of solutions generated could be one way forward.

It could also be emphasised that the use of a two-phase method, such as the 'locate-allocate' approach commonly used in multi-source Weber problem (or its discrete counterpart the p-median problem), has some similarities with HC where in Phase 1, p clusters are obtained and in Phase 2, the optimal 1-median problem is optimally solved. The new locations are then used to construct the new p clusters for Phase 1 and the process continues until there is no change in the location of any of the p facilities. This approach is originally proposed by Cooper (1964) and applied by several authors including, see, Gamal and Salhi (2003) and Luis et al. (2011).

Multi-level Decomposition

When the problem is large, we can either use directly a powerful approach to tackle the entire problem through an efficient metaheuristic or attempt to integrate exact method as part of the methodology. Here a recursive approach that selects a subproblem from the entire problem (choose $n' << n$ where n is the original problem size such as the number of customers and n' the number of customers in the subset). This approach has some similarities with Concentration Set (CS) except that each subproblem is now solved optimally or via a powerful heuristic leading to an optimal configuration with its selected attributes. Consider Λ_k to be the set of the selected configurations when solving the kth subproblem, $(k = 1, ..., K)$ with K representing the number of samples or subproblems tested. Let $\Lambda = \cup_{k=1}^{K} \Lambda_k$ be the set of promising attributes. In facility location, these can represent the promising facilities. The reformulated problem can now use the number of potential sites as $m' \leq |\Lambda|$ instead of $m = n$ and solve the problem still obviously with n customers which can be attempted with an exact method or via a powerful metaheuristic such as VNS or TS. This approach was successfully explored by Irawan et al. (2014) when solving very large p-median problems with n reaching 70,000 customers and $p = 10, \ldots, 100$. Note that this approach can be extended and adapted easily to tackle any large combinatorial problems where promising attributes could be identified when solving a subproblem. Possible extensions may include guiding the sampling by introducing some of these promising attributes as part of the solution when solving a given subproblem especially after a few iterations were tried or just by allowing the entire process to start again after K runs were performed. This is an exciting way of dealing with large instances which can have a lot of scope for improvement.

5.4   Data Mining Hybridisation Search

When there is a large number of data and the aim is to extract some interesting information either known a priory or completely unknown (patterns, clusters, relationships, classification), schemes that are inspired from heuristics and sometimes supported by statistical modelling and exact methods are worth the investigation. For example, in loan insurance or mortgages, customers get a yes or a no score to assess whether the customer is illegible for the loan or not. This solution is based on the customers past information (age, gender, salary, home ownership, etc.). This can be seen as a classification issue where the instance here the customer is put into a specific category (yes or no). A neural network or decision-tree-based approaches are usually used to provide the user with such an important decision which can be vital for the customer. In many cases, the problem is so large that there will be interesting features and patterns in the data that cannot be easily discovered but are still worth identifying. Note that if the size of the problem is that big, modelling through standard optimisation or statistical modelling using commercial software can easily become obsolete.

One approach which I think to be worth exploring can be outlined as follows:

> At the beginning of the search and as no information is available, we can start by taking a few samples randomly and perform some simple analysis using very quick selection rules. This leads to some attributes that could be worth recording. At this stage, the technique used is either optimisation-based or statistical and can be rather crude and fast. This is repeated for a certain number of rounds until T_1 say, where those individual results could be analysed using a simple heuristic leading to useful information to be selected. In the next round of sampling, given some of the useful information are already embedded in, a similar analysis is carried out again. The choice of the samples and their sizes can be identified by heuristic rules as well. At this point, the technique that will be used becomes slightly more powerful and the sampling strategies relatively more sophisticated. This process of gathering information and increasing the level of solution quality by adopting a slightly more powerful heuristic as the search progresses is repeated several times. The switching to a higher calibre heuristic can be either dynamically identified or set at fixed points, say $T_2,....., T_K$. Note that at this stage enough information is collected and the problem sizes become more manageable until T_K is reached where we can then adopt till the end a more powerful metaheuristic, even an exact method based on mathematical programming, DP or just pure statistical modelling if possible. The choice of K, the switching positions (i.e., the $T_j; j = 1, \ldots, K$), the number of samples within each strip and their sizes all contribute to the final solution leading to challenging and exciting questions that need answers. This is one of the methodologies that can be used efficiently in tackling this problem of data mining.

For example, when solving very large p-median problems, a similar but slightly simpler methodology using sampling, VNS and ILP formulation was presented with interesting results, see Irawan et al. (2014, 2016) for the p-median and the p-centre problems, respectively.

A general schematic representation that shows the reduction of the problem size and the degree of focus (solution quality) as the approach

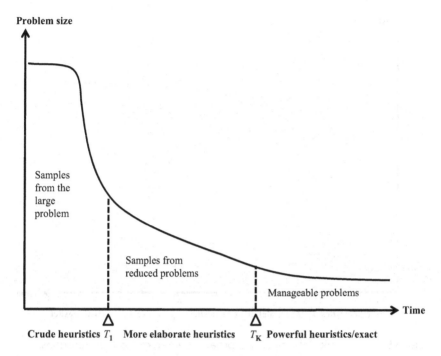

Fig. 5.4 Data mining hybridisation search heuristics

progresses with respect to time or iteration is shown in Figs. 5.4 and 5.5, respectively.

In summary, hybridisation is carried out using higher and higher level heuristics as the search progresses. In other words, crude heuristics or even simple rules are first used to perform the sampling as well as to analyse them, then followed by more advanced approaches such as composite heuristics or metaheuristics but applied within a small computing time or based on a reduced number of neighbourhoods, then finally a well-defined and structured problem could be defined which is then addressed in a standard optimisation or statistical manner where powerful metaheuristics or exact methods are adopted if possible. This type of hybridisation which is heavily recursive and time dependent of the progress of the search may not be that simple to explore but, in my view, is worth the challenge.

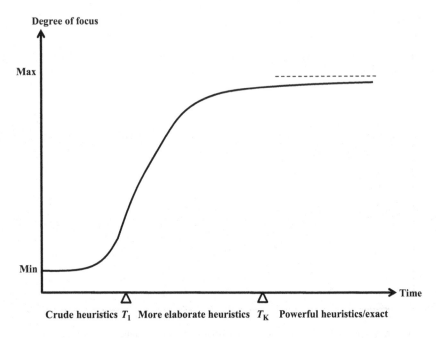

Fig. 5.5 A dynamic data mining approach (degree of focus vs iterations)

5.5 Summary

The power of an individual approach though can be very strong, taking advantage of the strengths of more than one approach when searching for a good quality solution is extremely powerful. This mechanism of intelligently integrating various methods leads to hybridisation. In this chapter, some well-known hybridisations that are based on hybridising heuristics with heuristics, heuristic within exact with heuristics and exact within heuristics. A special hybridisation devoted to data mining is also given. The design of efficient hybridisations will be challenging and worth the effort especially the matheuristic-based ones as they have the advantage in guaranteeing optimality or at least providing interesting bounds.

References

Agar, M., & Salhi, S. (1998). Lagrangean heuristics applied to a variety of large capacitated plant location problems. *The Journal of the Operational Research Society, 49,* 1072–1084.

Akpinar, S. (2014). Hybrid large neighbourhood search algorithm for capacitated vehicle routing problem. *Expert Systems with Applications, 61,* 28–38.

Angelelli, E., Mansini, R., & Speranza, M. G. (2010). Kernel Search: A general heuristic for the multi-dimensional knapsack problem. *Computers and Operations Research, 37,* 2017–2026.

Beasley, J. E. (1993). Lagrangean heuristics for location problems. *European Journal of Operational Research, 65,* 383–399.

Boudia, M., & Prins, C. (2009). A memetic algorithm with dynamic population management for an integrated production-distribution problem. *European Journal of Operational Research, 195,* 703–715.

Bouzid, M. C., Ait Haddadene, H., & Salhi, S. (2016). A new integration of Lagrangean split and VNS: The case of the capacitated vehicle routing problem. *Computers and Operations Research.* doi:10.1016/j.cor.2016.02.009.

Burke, E., Kendall, G., Newall, J., Hart, E., Ross, P., & Schulenburg, S. (2003). Hyper-heuristics: An emerging direction in modern search technology. In F. Glover & G. A. Kochenberger (Eds.), *Handbook of metaheuristics* (pp. 457–474). London: Springer.

Burke, E. K., Hyde, M., Kendall, G., Ochoa, G., Ozcan, E., & Woodward, J. (2010). A classification of hyper-heuristics approaches. In M. Gendreau & J. Y. Potvin (Eds.), *Handbook of metaheuristics* (pp. 449–468). London: Springer.

Chen, D., & Chen, R. (2009). New relaxation-based algorithms for the optimal solution of the continuous and discrete p-centre problems. *Computers and Operations Research, 36,* 1646–1655.

Cooper, L. (1964). Heuristic methods for location-allocation problem. *SIAM Review, 6,* 37–53.

De Franceschi, R., Fischetti, M., & Toth, P. (2006). A new ILP-based refinement heuristic for vehicle routing problems. *Mathematical Programming B, 105,* 471–499.

Elhossini, A., Areibi, S., & Dony, R. (2010). Strength Pareto particle swarm optimization and hybrid EA-PSO for multi-objective optimization. *Evolutionary Computation, 18,* 127–156.

Fischetti, M., & Lodi, A. (2003). Local branching. *Mathematical Programming, 98*, 23–47.

Fischetti, M., Glover, F., & Loti, A. (2005). The feasibility pump. *Mathematical Programming B, 104*, 91–104.

Gamal, M. D. H., & Salhi, S. (2003). A cellular heuristic for the multisource Weber problem. *Computers and Operations Research, 30*, 1609–1624.

Garcia-Villoria, A., Salhi, S., Corominas, A., & Pastor, R. (2011). Hyperheuristic approaches for the response time variability problem. *European Journal of Operational Research, 211*, 160–169.

Geoffrion, A. M. (1974). Lagrangean relaxation for integer programming. *Mathematical Programming Study, 2*, 82–114.

Guastaroba, G., & Speranza, M. G. (2012). Kernel search for the capacitated facility problem. *Journal of Heuristics, 18*, 877–917.

Hanafi, S., & Wilbaut, C. (2006). Mixed integer programming relaxation based Heuristics to solve the 0–1 multidimensional knapsack problem. *Paper given at COR/Optimization Days May 2006*, Montreal.

Hansen, P., Mladenović, N., & Urosević, D. (2006). Variable neighbourhood search and local branching. *Computers and Operations Research, 33*, 3034–3045.

Held, M., & Karp, R. M. (1970). The traveling salesman problem and minimum spanning trees. *Operations Research, 18*, 1138–1162.

Held, M., & Karp, R. M. (1971). The traveling salesman problem and minimum spanning trees: Part II. *Mathematical Programming, 1*, 6–25.

Irawan, C. A., Salhi, S., & Scaparra, P. (2014). An adaptive multiphase approach for large unconditional and conditional p-median problems. *European Journal of Operational Research, 237*, 590–605.

Irawan, C. A., Salhi, S., & Drezner, Z. (2016). Hybrid metaheuristics with VNS and exact methods: Application to large unconditional and conditional vertex p-centre problems. *Journal of Heuristics, 22*, 507–537.

Lazic, J., Hanafi, S., Mladenović, N., & Urosevic, D. (2010). Variable neighbourhood decomposition search for 0–1 mixed integer programs. *Computers and Operations Research, 37*, 1055–1067.

Luis, M., Salhi, S., & Gabor, N. (2011). A guided reactive GRASP for the capacitated Multi-source Weber problem. *Computers and Operations Research, 38*, 1014–1024.

Mashwani, W. K., & Salhi, A. (2010). Multiobjective memetic algorithm based on decomposition. *Applied Soft Computing, 21*, 221–243.

Moscato, P. (1999). Memetic algorithms. In D. Corne, M. Dorigo, & F. Glover (Eds.), *New ideas in optimization* (pp. 219–235). New York: McGraw Hill.

Moscato, P., & Cotta, C. (2003). A gentle introduction of memetic algorithms. In F. Glover & G. A. Kochenberger (Eds.), *Handbook of metaheuristics* (pp. 105–144). London: Kluwer.

Moscato, P., & Norman, M. G. (1992). A memetic approach for the travelling salesman problem- implementation of a computational ecology for combinatorial optimization on message passing systems. In M. Valero et al. (Eds.), *Parallel computing and transputer applications* (pp. 177–186). Amsterdam: IOS Press.

Norman, N., & Iba, H. (2008). Accelerating differential evolution using an adaptive local search. *IEEE Transactions on Evolutionary Computation, 12*, 107–125.

Nwana, V., Darby-Dowman, K., & Mitra, G. (2005). A co-operative parallel heuristic for mixed zero-one linear programming: Combining simulated annealing with branch and bound. *European Journal of Operational Research, 164*, 12–23.

Raidl, G. R. (2006). A unified view on hybrid metaheuristics. In *Hybrid metaheuristics* (pp. 1–12). Berlin/Heidelberg: Springer.

Rosing, K. E., & ReVelle, C. S. (1997). Heuristic Concentration: Two stage solution construction. *European Journal of Operational Research, 97*, 75–86.

Rosing, K. E., ReVelle, C. S., & Schilling, D. A. (1999). A Gamma heuritsic for the p-median problem. *European Journal of Operational Research, 117*, 522–532.

Ross, P. (2005). Hyper-heuristics. In E. K. Burke & G. Kendall (Eds.), *Search methodologies: Introductory tutorials in optimization and decision support techniques* (pp. 529–556). London: Springer.

Salhi, S. (2006). Heuristic search in action: The science of tomorrow. In S. Salhi (Ed.), *OR48 keynote papers*, ORS Bath, pp. 39–58.

Salhi, S., & Al-Khedhairi, A. (2010). Integrating heuristic information into exact methods: The case of the vertex p-centre problem. *The Journal of the Operational Research Society, 61*, 1619–1631.

Salhi, S., & Sari, M. (1997). A multi-level composite heuristic for the multi-depot vehicle fleet mix problem. *European Journal of Operational Research, 103*, 78–95.

Sorensen, K., & Sevaux, M. (2006). MA/PM: Memetic algorithms with population management. *Computers and Operations Research, 33*, 1214–1225.

Talbi, E. G. (2016). Combining metaheuristics with mathematical programming, constraint programming and machine learning. *Annals of Operations Research, 240,* 171–215.

Thangiah, S. R., & Salhi, S. (2001). Genetic clustering: An adaptive heuristic for the multi depot vehicle routing problem. *Applied Artificial Intelligence, 15,* 361–383.

Wilbaut, C., Hanafi, S., Freville, A., & Balev, S. (2006). Tabu search: Global intensification using dynamic programming. *Journal of Control and Cybernetics, 35,* 579–598.

Wilbaut, C., Hanafi, S., & Salhi, S. (2008). A survey of effective heuristics and their application to a variety of knapsack problems. *IMA Journal of Management Mathematics, 19,* 227–244.

Wilbaut, C., Salhi, S., & Hanafi, S. (2009). An iterative variable-based fixation heuristic for the 0–1 multidimensional knapsack problem. *European Journal of Operational Research, 199,* 339–348.

Xiao, Y., Zhao, Q., Kaku, I., & Mladenovic, N. (2014). Variable neighbourhood simulated annealing algorithm for capacitated vehicle routing problems. *Engineering Optimization, 46,* 562–579.

6

Implementation Issues

6.1 Data Structure

Since metaheuristics are computing intensive (a large number of iterations are usually performed), any intelligent way of reducing the computer time without significantly affecting the solution quality. Such a saving in computer time can be reallocated if necessary to perform additional iterations which may lead to even better quality solutions. Some ideas that may help in emphasising this concept of DS while writing a computer program are outlined below:

Avoidance of the Recomputation of Already Computed Information

This can be achieved by storing the necessary *intermediate* information (these data will usually be not kept and hence not taken advantage of if DS is not used). There will be a fixed cost of designing such a DS besides the additional memory that this may require. But nevertheless this is worth especially if the problem requires a lot of iterations as usually the case in metaheuristics.

© The Author(s) 2017
S. Salhi, *Heuristic Search*, DOI 10.1007/978-3-319-49355-8_6

For instance, consider the case of the vehicle routing problem; we have NR vehicle routes denoted by $(R_1, \ldots, R_{NR})$ and we are performing the best customer insertion into these routes. Initially, for each route, say R_i, all customers not in R_i are tested for possible insertion and the customer that maintains feasibility and yields the cheapest insertion cost is chosen. The best overall insertion is then calculated over all the best insertions within each route. Suppose the current solution is found by inserting a customer from R_k into R_s. When looking for the next neighbouring solution, we do not need to perform once again all the calculations. Since we know the best insertion within routes (if we have kept such an information!) all what we need to check are those customers from R_s and R_k to be inserted in other routes and customers from other routes to be inserted into R_k and R_s. In other words, we only recompute the information where the changes occurred (i.e., we concentrate on those affected routes only as the other information is still unchanged). Such a DS was successfully adopted in the vehicle routing by Osman (1993) and also in the vehicle routing with different capacities by Osman and Salhi (1996). This reduced the number of iterations from $\frac{NR(NR-1)}{2}$ to $2(NR-1)$ each time the best insertion of a customer is needed which can be massive. The question here is how to identify systematically the new affected routes without recourse to checking all combinations and so on. The idea here is to first determine the best improvement (which may not be necessarily positive) obtained when swapping customers or exchanging customers between two routes, their respective insertion places and any other attributes that are worth recording. This information needs to be stored efficiently and its retrieval access made very easy. In this case, consider two matrices A and B where A has a dimension $NR \times NR$ with the upper diagonal part recording the improvement $\Delta_{rs} (r, s = 1, \ldots, NR)$, the diagonal elements store the $\Delta_{rr} (r = 1, \ldots, NR)$ representing the improvement due to the intra route moves, and the lower triangular part $I_{rs} (r < s)$ representing the positional index which is defined as $I_{rs} = r + ((s-1)(s-2)/2); s < r$. The matrix B has dimension $N \times M$ with $N = NR(NR-1)/2$ and $M = \#$ attributes with the first column referring to the position index. The other attributes refer to affected routes, the best insertion of the exchanged customers and type of operators chosen for such best improvement. For instance, if the best improvement is found by $\Delta_{rs} = \text{Max}\left(\Delta_{jk}; j, k = 1, \ldots, NR\right)$, this shows that position index is I_{rs} leading to the attributes in $B(I_{rs}, l), l = 1, \ldots,$ M being identified easily. For instance, $B\left(I_{kj}, 2\right) = r$ and $B\left(I_{kj}, 3\right) = s$, etc. Routes r and s are affected and all the operators will be performed using the combinations that involve these two routes such as $(r, k), (s, k), (r, s); k = 1, .., NR; k \neq r, s$. Other information could also be added in other matrices if necessary so the retrieval can be performed quickly. For more information on this construction, see the recent work by Sze et al. (2016) and references therein.

The above example is used for illustration only. In most combinatorial optimisation problems, various local searches are usually used intensively and therefore recording intermediate information are useful, and hence constructing a DS that reflects that is paramount. Though there would be an initial fixed cost to bear and a small computational burden linked to recording and updating the information, this effort is relatively negligible considering the large number of iterations used in metaheuristics-based methods. The advantage of such DS is that the solution quality is not affected as there is no need for compromising between solution quality and computation time required which leads to a win-win situation.

Elimination of Common Tasks Which Are Used in Several Parts of the Method

For instance, the cost (total demand or time) of a vehicle route without a given customer remains unchanged from one iteration to the next. This information though small helps enormously in cutting redundant calculations as shown by Salhi and Sari (1997) when solving the vehicle routing with heterogeneous vehicle fleet. This obviously requires an additional storage requirement but this is usually not a severe restriction. This clue is worth incorporating in other applications where intermediate information could be identified.

Avoidance of Extra Computation

This can performed by detecting as early as possible the possible conditions or restrictions that may not permit a given move to be worth evaluating. For instance, in any constrained type problem, say the vehicle routing problem, if a given customer is introduced into a vehicle route which violates capacity or time constraints, it would be economical if such an information was made available as early as possible in the search so to cut down on any earlier worthless computation. We assume we base our search on feasible moves only, otherwise extra care is needed as the amount of violation can be used as a target to whether or not it is worth exploring such moves. Another example is the vertex p-centre problem where the aim is to locate p facilities among the existing customer sites such that the furthest customer from its nearest chosen facility is minimised. This problem relates to emergency services such as the case of locating hospitals, fire service or police stations. One of the approaches is based on reducing the gap between an LB and UB (thresholds or response times) by applying a bisection method in the range to define the new threshold for which a set covering problem is applied. This is repeated until there is nothing left in the range. One interesting view is to sort the distances between nodes and therefore identify quickly those distance elements that lie in a range (between the current LB and UB) at any iteration without considering the others which will be many. The bisection point, which will be used as the next threshold, is then computed but chosen as the nearest existing distance element in that range and not as its original value which may not be necessary one of the elements in the range. Besides, if there is only 1 or 2 elements left in the range even if the range is still wide, a complete enumeration of these elements used as thresholds can be conducted instead which can be much faster. The search then terminates much quicker without searching for bisection point all the way until there is a range of size one. Irawan et al. (2016) incorporated this scheme within their hybrid exact/VNS framework to produce optimal solutions for large instances for both the p-median and the vertex p-centre problems for the first time.

Tree-Based DS

One well-known data compression technique is the quadtree-based DS commonly used in computer science. This was successfully adopted by Salhi and Irawan (2015) to solve very large p-median problems. Here, the data are first classified under quadtrees which are well defined by their corners. The idea is to construct a tree starting from the first node representing the overall region, which is then split into four branches (the initial space is split into four equal spaces, usually rectangles); this splitting continues until the number of lowest level is reached. This mechanism requires a high fixed cost at the beginning of the search which is then compensated by the massive saving in later stages of the search. The idea is that once the four corners of each quadrant are assigned to their nearest facility, all those nodes (customers) inside that quadtrees are systematically assigned to that facility as one block instead of individual nodes resulting in a massive saving in computational time. This DS is practically useful if the size of the problem is very large and many runs are required as the construction of the DS though time consuming is performed once only.

A simpler version that also uses tree splitting is hierarchical clustering commonly used in statistical applications. Here a nested sequence of partitions forming a hierarchal tree is based on splitting each successive layer of the data of the above group into two smaller groups instead. Following the same principle, B&B trees, mainly used for integer linear programming problems, are also formed in the same way. The storage of information at each node of the tree is critical for subsequent trees.

In brief, the design of an effective DS is useful as a way to keep track of the already found information. In other words, there is <u>no need to recompute</u> wholly or partially already computed operations if these were efficiently stored and easy to retrieve. This can save a large amount of computing time that can be used for additional tasks if need be.

6.2   Duplication Identification

In addition to the basic DSs, some of which can be problem specific, there are several ways on how to identify duplicate solution configurations during the search. These mechanisms are introduced to avoid the risk of cycling as well as the unnecessary recording of similar already found information that do not need to be examined again. Some of these schemes are presented here.

Hashing Function

A hashing function is one way to efficiently record and detect previously found solutions, see Woodruff and Zemel (1993). These functions which are defined from the solution configuration set need to be assigned integer numbers while being bijective (i.e., a one-to-one relationship). In other words, each value of the hashing function will relate to one single solution configuration only. The construction of such functions is not easy as these functions need to be theoretically proven to guarantee the unique interdependency between the hashing function value and the corresponding configuration. In practice, a nearly bijective function is usually adopted as a proxy with a small degree of inaccuracy.

For instance, in vehicle routing, possible functions may include

(i) the objective function value, namely the total cost $Cost(S)$,

(ii) some index function based on route configuration
$$H(S) = \sum_{j=1}^{NR} \sum_{i=1}^{NC_j} \sigma_{ij} 2^i \text{ where } \sigma_{ij} \text{ refers to the } i^{th} (i = 1, \ldots, NC_j)$$
customer in route $j (j = 1, \ldots, NR)$ with NR denoting the number of routes as before and NC_j the number of customers in route j. The index i refers to the position in the route configuration, or

(iii) a slightly simpler form than (b) is $H(S) = \sum_{j=1}^{NR} NC_j \sum_{i=1}^{NC_j} \sigma_{ij}$.

Simultaneous Use of Multiple Simple Performance Functions

However, the issue can be made slightly less difficult by defining a given configuration using not necessarily one single function but with a few functions which when these are put together will reduce the risk of missing the 'one-to-one relationship' property. For instance, in the case of routing problems, we can use the following checks in sequence. We can conclude that if the total cost is different, then the new solution is different, otherwise, we check if the number of routes is different, then if the number of customers in each route is different, if the load in each route and so on.

String-Based Checks

Another way to identify solution duplication is simply to transform the sequence of each route by a string and just compare the strings. The latter is found to be effective when combined with others such as the checking of the objective function value, for example.

Set-Based Identifications

The issue of duplication is relatively easier to deal with in those applications where the solution configuration is defined by a selected subset of integer numbers instead of a sequence. For instance, in the p-median problem where the configuration is made up of p integer numbers denoting the site number of the facilities, we can sort each configuration in ascending (or descending) order of its sites numbers as the order has no practical meaning here. If the total cost (or the objective function value of the problem studied) happens to be the same as some of the previous configurations, the following two tricks, which are shown to be efficient, include (i) redefining the ordered the subset into a string and apply (c) above, or (ii) perform the checks column by column only starting from the first one. For example, if the kth position ($k = 1, \ldots, p$) is different to the kth position of the previous configuration, the new configuration must be different and therefore there is no need to check the remaining positions. This overall check of the new configuration, if m other already found configurations have the same cost, will take $0(pm)$ in the worst case only.

6.3 Approximation Effect on Cost Function Evaluation

The computation of the objective function, once the move is selected, can require excessive computational effort. As there will be a lot of moves to evaluate, if such an evaluation is too heavy computationally, it could be important to apply a good but quick approximation of the cost evaluation instead. To add flexibility and reduce the risk of missing good solutions, the top K moves will be rechecked properly using the full cost evaluation. Note that if K represents all the possible moves, this will reduce to the original implementation but if $K = 1$, the chosen move will be based on the best one using the approximation. The former could be too time consuming, whereas the latter is quick but could be too myopic, and hence could miss good quality moves. Having K a constant, say $K = 10$, a fraction of the problem size or just based on the number of possible moves could be a good way forward. Such implementation is proved to be successful in solving capacitated location problem where the full TP is solved optimally using the selected moves only and the approximation is based on assigning customers to the open facilities using the nearest proximity criterion or the regret cost (difference between the second nearest and the nearest facility to a customer) instead. The main idea is that the approximation needs to be easy to compute while being as close as possible to the real objective function and robust in the sense that all moves are similarly approximated irrespective whether they are all over or under estimated.

6.4   Reduction Tests/Neighbourhood Reduction

The reduction in computing time can be obtained by introducing some neighbourhood reductions which eliminate from the testing certain cases are <u>unlikely to influence</u> the global best solution. There will be a trade-off between speed by which the best solution is obtained and the quality of the solution. When implementing a heuristic, it is important to understand the problem which will help in defining the right size of the neighbourhood for each attribute. This mechanism, also known as a reduction test, can be built to be either dynamic or deterministic set from the outset. For instance, Salhi and Sari (1997) introduced some reduction tests when studying the multi-depot vehicle routing and saved over 85 % of the CPU time with a negligible loss in solution quality, some of which are based on the single depot, whereas others rely on the structure of neighbouring depots. A good neighbourhood reduction scheme has also the added benefits in matheuristics in some circumstances, as it transforms the original problem into a new reduced problem that could be manageable to be optimally solvable as discussed in the previous chapter. A similar approach that considers the power of neighbourhood reduction is the interesting paper on granular TS by Toth and Vigo (2003) where restricted neighbourhoods are defined making the overall search relatively much faster. It is worth mentioning that the use of neighbourhood reduction does not guarantee the same level of solution quality though the challenge is to construct such schemes which only exclude with high probability those unnecessary moves. This risk presents an exciting challenge with the aim to thrive towards finding the right balance between a strong reduction test (remove as much as possible) while not affecting the solution quality. Note also that, in very rare cases due to adopting heuristics instead of an exact method, the new solution found could even be better based on the same number of iterations. However, when the same CPU time is used as the stopping criterion instead, the new solution could easily improve on the original one as many more iterations will be tried leading to more moves being evaluated.

For illustration purposes, I shall present some of those neighbourhood reductions that are originally proposed for the vehicle routing by Salhi and Sari (1997). These can be extended and refined if necessary. For instance, very recently these were explored further and used successfully in solving large vehicle routing problems by Sze et al. (2016). Neighbourhood reduction schemes for other topics do exist, sometimes the more constrained the problem is the more opportunities exist for powerful reduction schemes to be developed as all those combinations that are not worth evaluating are excluded from the computation of their respective moves.

Proximity-Based Scheme (Rule 1)

For each node $i \in \{1, \ldots, n\}$, let us define the subset of nodes that should not be excluded for subsequent moves where node i is involved, say $N(i)$. This can be defined as $N(i) = \left\{ j \in \{1, \ldots, n\} : d_{ij} < \widetilde{D}_i \right\}$ with

$$\widetilde{D}_i = k.\frac{\sum\limits_{j=1}^{n} d_{ij}}{n-1}$$ and $k \leq 1$. A simpler version is to choose for each node $i \in \{1, \ldots, n\}$ the K nearest nodes with $K_0 = \left\lceil \frac{n}{LB} \right\rceil$ with LB being the LB on the number of vehicles.

This is defined as $LB = \left\lceil \dfrac{\sum\limits_{j} q_j}{VC} \right\rceil$ with VC being the vehicle

capacity and q_j the demand of node $j \in \{1, \ldots, n\}$.

To strengthen the above rule and make it node-based, K_0 can be extended to $K_i = \text{Max}\left\{ K_{\min}, \text{Min}\left(K_0, |N(i)|\right) \right\}$ with $K_{\min}$ set to a small value to guarantee that a small number of nodes is at least evaluated, say $K_{\min} = 5$.

In summary, for each $i \in \{1, \ldots, n\}$, $N(i)$ is first defined and the following logical flag is set as: $\text{Flag}(i,j) = \begin{cases} \text{true} & \text{if } j \in N(i) \\ \text{false} & \text{Otherwise} \end{cases}$

Initially all Flag(i, j) are set to false and updated accordingly. If no reduction scheme is used, this is equivalent to setting all Flag(i, j) to true.

Relaxing Flag(i, j) Due to Depot Location (Rule 2)

Some customers may be far away from each other but could be close to the trajectory of leaving the depot or returning to the depot. This observation can be taken into account by introducing the following additional reduction test.

For each $i \in \{1, \ldots, n\}$, let the insertion cost of node j between the depot, say 0, and node i be as follows: $I(i, j, 0) = d_{ij} + d_{oj} - d_{0i}$

For $i \in \{1, \ldots, n\}$ and   for   each $j \notin N(i)$ set

$$\text{Flag}(i, j) = \begin{cases} \text{true} & \text{if } I(i, j, 0) \leq \widehat{I}_i \\ \text{false} & \text{Otherwise} \end{cases}$$

where $\widehat{I}_i = \dfrac{\displaystyle\sum_{j \in N(0)} I(i, j, 0)}{|N(0)|}$ and $N(0)$ defined similarly as N(i).

The above two rules are represented in Fig. 6.1 for an example of two routes. In the case of rule 1, for simplicity, two circles based on two particular nodes, one from each circle, are shown. Note that there will be many circles, one for each node i representing $N(i)$. Here, we can see that node j will not be considered for any move evaluation that involves i or any nodes in those two circles. However, in rule 2, node j seems to be on the trajectory towards the depot based on a small extra cost due to its insertion making it worth exploring. Note that though node j will be considered for moves with node i, it is still not worth being involved with the other nodes in those circles.

Angle-Based Scheme (Rule 3)

Another test that considers those pairs of nodes (i, j) which happen to be within a certain angle with respect to the depot, $\theta(i, 0, j)$, or not too far apart but within proximity to the depot and the two nodes is also worth adding into the overall pruning mechanism in any move evaluation that

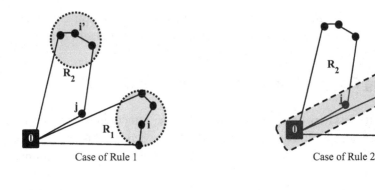

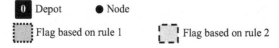

Fig. 6.1 Two neighbourhood reductions for the VRP

uses a pair of nodes. In other words, if $\text{Flag}(i,j) = \text{false}$ due to the previous two rules but

if $\{\theta(i,0,j) \leq \theta_{\min} \text{ (say } \pi/12)\}$ or $\{\theta(i,0,j) \leq \theta_{\max} \text{ (say } \theta_{\max} = 2 \theta_{\min}; \text{ eg., } \pi/6) \text{ and both } i,j \in N(0)\}$ then $\text{Flag}(i,j) = \text{true}$.

A Simple Implementation and Adaptation

The above rules, or similar ones, are defined a priory, before the search starts using the following simple setting:

For all $i = 1, ..., n-1; j = i+1, ..., n$ set

$$\text{Flag}(i,j) = \begin{cases} \text{true} & \text{if } j \text{ satisfies any of the above rules} \\ \text{false} & \text{otherwise} \end{cases}$$

and $\text{Flag}(j,i) = \text{Flag}(i,j)$.

Another simple trick, which will not add any editing or amendment to those moves already successfully working under the environment that does not incorporate neighbourhood reduction, is to add at the beginning of each move, where the pair of nodes are involved, this simple check: {If (Flag(i,j) = false) next}. In other words, the search jumps to the next node j if the current pair (i,j) is not worth exploring as it fails the test.

The above rules are also implemented for other related VRP applications such as the multi-depot case. Here, the borderline nodes (those nodes that lie between their nearest and second nearest depots) use the above three rules with respect to their first and second depots. However, those non-borderline nodes (i.e., the rest of the customers) are assigned to their respective depot (i.e., nearest) using the above three rules within their respective clusters.

As a little side story, when I implemented the above methodology (see Salhi and Sari 1997) for the multi-depot routing problem on a beautiful spring day in 1990, I initially tried it for the largest instance at that time which used to take just over 1 hour (64.5 minutes) in a Sun Sparc station 10. After adding the above tricks which I call 'reduction tests', also known by their more sophisticated name 'neighbourhood reduction', I had a shock as the search terminates just over 3 minutes (3.3 minutes to be precise!). I thought there is a mistake somewhere and the search must have skipped everything. When I checked the full results and to my surprise the solution including the vehicle routes and so on were exactly the same for that particular instance, I was confused, it is like winning a jackpot! I knew mathematically that the search will be much faster due to the elimination of all those redundant calculations, but I did not imagine it to be that fast!

6.5   Parameters Estimation/Adaptive Search

One of the weakest and most criticised aspect within heuristics is usually the reliance of good estimate and fine tuning of the parameters used. There are usually four ways to select the parameters:

(a) *Experimental work* This is one of the most used techniques which aid to find the most appropriate values after performing preliminary testing usually conducted on a small set of instances.
(b) *Analytical forms* This aims to determine analytically these values which can be derived statistically via regression, for example.

(c) *Matching methods* The idea is to assign the parameters already found by other people to a given class of problems.

(d) *Adaptive adjustment* This is based on dynamically adjusting the parameters values depending on the solution quality as the search progresses.

The second approach (b) is usually the most exciting one though it can be difficult to define an analytical function that works all the time. For instance, Salhi (2002) put forward a formulae for defining tabu length and aspiration level within TS, and Dorigo et al. (1999) gave a formula for the number of ants based on the other parameters within ant systems. An interesting statistical investigation by Adenso-Diaz and Laguna (2006) was also conducted for a metaheuristic which is applicable for other ones. An earlier study focussing on GA was also attempted by Grefenstette (1986). The fourth category (d) is the one that starts to gain more attention as it shows to be robust besides yielding more competitive results when compared to static measures. Here, some form of control is embedded into the search so corrective measures are added to either avoid the risk of early convergence or to control the problem of divergence. It is also important for the search to be able to restrict some parameters to fixed values based on the problem characteristics and the history of the solution at a given point of time. This guidance will rely on a smaller set of parameters instead but it ought to have some backtracking facilities if necessary.

The use of information to control the search is successfully applied in several applications. As examples, consider the RTS where the tabu size goes up or down depending on the solution quality and the risk of collision. In SA, the temperature is reset to previously found values, in VNS, for larger and complex problems, the use of a smaller number of promising neighbourhoods is restricted in pseudo-random way to a smaller but promising subset among the original set of neighbourhood structures. The same idea is applied to the use of a subset of local searches instead of the full set which may require an excessive computational effort. The choice of such subsets is performed dynamically, that is, based on those good solutions found earlier in the search. In GA, the rate of selection between the elite solutions and the less well-off ones is also controlled dynamically as the search progresses.

Neural Network One approach which relies heavily on dynamic adjustment is the construction of the weights in an artificial neural network. Here, the weights between the layers are adjusted recursively until a steady state of appropriate weights is identified, usually based on the minimisation of the sum of square of the errors. This can be applied in two ways either as a black box where a new entry details (e.g., the input of a customer requesting a loan) is entered into the network which is already well defined, and hence the decision comes as either accept or reject the applicant. However, the black box could also have the facility to incorporate new entries and fine tune the network even better. The standard architecture for supervised learning is the well-known multi-layer feed forward network studied by Hopfield and Tank (1985), and Rumelhart and McClelland (1986). The network consists of input, hidden and output layers with weights representing the importance of the connections between two nodes. Based on a set of given input patterns and their corresponding target outputs, the training is applied to minimise the sum of the errors between the unknown output patterns and the known target patterns. This problems turns out to be an optimisation problem with the weights as the decision variables. Initially, the problem is solved by back propagation with a decent method as its optimiser but there are more powerful optimisers including adaptive TS as given by Smithies et al. (2004).

6.6 Constraint Handling via SO

Metaheuristics in general incorporate infeasible solutions in their search to widen the solution space. This is achieved by constructing an augmented objective function made up of the original function and the penalty cost due to infeasibility which is measured by attaching a penalty value, p, to the amount of violation. This concept is widely used in constrained non-linear optimisation, LRH and dual ascent methods as applied to linear programming.

The process of SO can be briefly outlined below but the reader can find more information in Kelly et al. (1993) and Gendreau et al. (1994):

- Initialise p close to zero, say $p = \delta$ (say $\delta = 0.1$)
- if the best solution is not improved set $p = p + \delta$ (after a number of iterations, say $K = 100$)
- once a feasible solution is found (say with $\widetilde{p}$), p oscillates between $[\widetilde{p} - M, ..., \widetilde{p} + M]$ say $M = 9$ or 10.

SO has three main advantages:

(i) It provides a mechanism for crossing regions of infeasibility during the search to reach a near-optimal solution. This happens as the whole feasible region (a) is made up of disconnected feasible regions where it is impossible to reach the optimal solution by moving from one region to another, (b) though the feasible region is connected, it is non-convex. If the search was constrained to move from a feasible solution to another feasible solution, it would be impossible to cross the boundaries, as in case (a) if the solution happens to be in the less attractive region, and in case (b), it may be tedious and long winded search to reach the good solution. In brief, allowing crossing between feasible and infeasible regions provides an efficient way to reach a good local optimum or near-optimal solution.

(ii) It is a flexible technique that can be used to locate high-quality feasible solution. It has some similarities with the way LRH works except that the latter provides a direct relationship between the feasible and the infeasible problems through the dual gap which is obviously important for heuristic evaluation purposes.

(iii) It has an indirect effect on any heuristic search that aspires long-term memory. Here, SO increases the heuristic power through the use of a non-dynamic objective function evaluation criterion. The penalty for infeasibility is very small at the start of the search and then increases either by a constant step size (δ), say $p = p + \delta$ or by a constant rate (α) as $p = \alpha p$ (say $\alpha = 2$) until a feasible solution is found. Once the feasible solution is obtained, the value of p changes linearly but oscillates around the value $\tilde{p}$ for which feasibility is currently found. In such circumstances, p may need to be decreased to give emphasis on feasibility and the whole process continues.

The way the penalty p is defined as the search progresses can be crucial as some guidance is required to stop the search from completely diverging. In some cases, imposing a maximum violation besides the updating of p is also used to guarantee to retain the search not too far away from feasible regions as it may become difficult otherwise to revert back to such feasible solutions.

6.7   Impact of Parallelisation

There exist various models on how to parallelise a given algorithm including master slave, fine grained, island and obviously a hybridisation of the above. The first one is to use several computers (or processors) and carry out a scheduling task between them, whereas the others especially the island model focuses on exploiting the power of course grain parallel computers. Even a computer with one processor only which happens to have dual or multi-core (internal parallel processing) can be taken advantage of its architecture. In addition, parallel random access model (PRAM) could also make use of memory better and hence activate the processing accordingly. Very recently, Graphics Processing Units (GPU), which is originally used in the video game industry, can be explored.

Master-Slaves Model

Most metaheuristics, especially population-based algorithms, are suitable methods for being implemented in parallel. The best move to be selected among those neighbouring solutions can be detected relatively quickly if the entire task is split into smaller tasks where the best intermediate solution will be found at each processor level. The overall best solution is then chosen using one of the processor or the master processor. A basic illustration of parallelisation of a search is given in Fig. 6.2 where there are

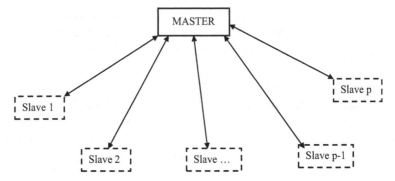

Fig. 6.2 Exchange of information between the master and the slave processors

p processors, called slaves, and the main one referred to as the master. The slaves perform operations and transfer the results to the master which gathers the best results from each slave. It then performs some operations, stores necessary information and sends back other tasks to the slave processors to perform. Note that the latter processor could also act as one of the slaves but extra care will be needed as the tasks required by the master processor are usually performed only when the other slaves have all completed their individual tasks. This process continues until the search is terminated. The saving in computing time although important is not $\frac{1}{p}$ of the original sequential time as one may expect. The saving is defined as follows.

Speed of Parallel Processing

The ideal speed-up (isu) is: $isu = \frac{T_n + T_u}{\left(\frac{T_n}{p}\right) + T_u} = \frac{T_n + T_u}{T_p + T_u}$

where

p is the number of processors

T_n is the sequential time to perform the overall computation if not parallelised

T_u is the time associated with the non-parallelisable part of the algorithm

T_p is the time parallelised when using p processors

In practice, $T_p \geq \frac{T_n}{p}$ due to the scheduling of the p tasks which is restricted by the time of the processor requiring the longest task (i.e., the bottleneck effect).

The exchange of information between the master and the slave processes needs to be planned carefully. Organising the parallel jobs efficiently can enhance the algorithm and get a speed-up close to isu. This problem is a job shop scheduling where there are p machines interrelated by the master machine. The aim is to balance the work load the slave processes while minimising the delay caused by any of the processors as this creates a bottleneck which will slow the whole process. This is a scheduling problem which is commonly studied in the literature, see Parhani (2006) for an overview on this subject.

As an example, GA by nature is among those metaheuristics that fits nicely into this class of methods which can be easily parallelised. For instance, when applying the GA, the computation of the decoding of the chromosomes as well as the fitness function values can be carried out in parallel, where each processor is assigned to a given chromosome (or a subset of chromosomes) to evaluate. In this case, each slave processor carries out the evaluation of its assigned part and sends its results to the master processor. This slave processor is then systematically assigned to the next available part that is not yet been evaluated. This process is repeated until the entire population is checked. The master processor then carries out the strategic operations such as the sorting of the chromosomes based on their fitness values, the removal of the unfit chromosomes, and any other decisions that make up the new population. Note that this sorting could also be performed relatively quicker if temporary sorting is performed. This can be achieved by performing the master task while the other slaves are conducting their jobs. In other words, every time a chromosome is evaluated, its fitness value is inserted in the right place according to the sorting criterion used instead of waiting till the end when all chromosomes are evaluated. This parallelism will undoubtedly cut down on the computer time considerably while maintaining the solution quality intact.

This process is easily adaptable for other population-based algorithms discussed in Chap. 4 such as ants systems, bees colony, particle swarm, harmony search and so on. Here, the pheromone trails for ACO, the velocities for PSO, the harmonies for HS and so on can be evaluated individually or in small groups, as in the GA, by the slave processors and the master processor decides how to update their corresponding values, how to choose the next move and whether or not to terminate the search.

Internal Parallel Processing

Recent PCs have multiple cores, usually two or four, to internally perform parallel processing if a call to such a function is activated. However, the code needs to be obviously written in a structure that can be parallelised. For instance, if we aim to find the maximum element of a matrix $(C_{ij})_{i,j=1,\ldots,n}$, say $CMAX$, its mathematical compact form can be written as $CMAX = \mathrm{ArgMax}\{C_{ij}; i,j = 1, \ldots, n\}$ and its corresponding code say in C++ could be the following:

$$CMAX = C[0][0]$$
$$\mathrm{For}\,(i = 0; i < n; i + +)$$
$$\{$$
$$\mathrm{For}\,(j = 0; j < n; j + +)$$
$$\{$$
$$\mathrm{If}\,(C[i][j] > CMAX)\,CMAX = C[i][j]$$
$$\}$$
$$\}$$

As you can see the maximum value is updated in a sequential way which will not be suitable for parallelisation even if the computer is equipped with multiple cores. Assume the computer has q cores. If we can take advantage of this computer facility, we can now reorganise the code in such a way that the q cores can be used more effectively by working in parallel. This task would not have been possible in the first implementation.

One way is to find for instance for each row i its corresponding maximum say $(CM_i)_{i=1,\ldots,n}$. Mathematically this is expressed as $CMAX = \mathrm{Max}\{CM_i; i = 1, \ldots, n\}$ such that $CM_i = \mathrm{Max}\{C_{ij}; j = 1, \ldots, n\}$

for all $i, i = 1, \ldots, n$ This mathematical form, though it is equivalent to the first compact one, can be easily translated into the following simple code which will be suitable to take advantage of this speed-up function.

```
For (i = 0; i < n; i + +)
{
CM[i] = C[i][0]
  For (j = 0; j < n; j + +)
  {
    If(C[i][j] > CM[i]) CM[i] = C[i][j]
  }
}
CMAX = CM[0]
For (i = 1; i < n; i + +)
{
  If(CM[i] > CMAX) CMAX = CM[i]
}
```

Both codes mathematically lead to the same value of $CMAX$. However, in the second code, the computer, if equipped with multiple cores, could perform q of those $CM(i)$ at the same time saving nearly $(q - 1)$ operations which is then repeated for the next q and so on (i.e., $\left\lceil \frac{n}{q} \right\rceil^{+}$ passes) until all the n rows have been examined leading to n new elements $CM(i); \ i = 1, \ldots, n$. To find $CMAX$ we need to find the maximum over this new vector $CM(i)$. At this stage, we can again take advantage of the multi-core processing by splitting this vector into smaller vectors of dimension at most $m = \left\lceil \frac{n}{q} \right\rceil^{+}$ or equivalent, where each part is individually processed by a given core and the maximum out of these is identified. As you can see though we need small additional operations at the end to have the maximum of these $CM(i)$ this is a relatively negligible task. It is worth noting that a longer code, which may not look relatively neater and concise than its initial counterpart, will be able to exploit this structure and be relatively more efficient in terms of computational speed. This is usually the case in coding as intermediate information could be stored and used whenever required. A similar observation can be noted when designing a DS that adds extra code and memory, as mentioned in an earlier subsection, but yielding a massive saving in computation time.

This aspect can be further enhanced by going deeper using the divide and conquer principle when finding the maximum of the $CM(i)$ or in the determination of the maximum within each row (i.e., computation of CM (i) itself). This is based on a better use of memory known as parallel random access model (PRAM). Here, successive pairs are first sorted and stored accordingly at odd positions leading to $\frac{n}{2}$ elements, this is repeated again yielding only $\frac{n}{4}$ elements where this process continues until there is one pair left only totalling $0(\log(n))$ iterations. This needs to be adjusted accordingly if necessary to cater for q cores or p processors.

Effect of GPU

GPU also called visual processing unit (VPU) are new concepts originally designed for video games and are available in most personnel computers. GPU handles thousands of cores, whereas CPU has only multiple cores. However, it is worth noting that each GPU core runs significantly much slower than its counterpart CPU core. The strength of GPU is for those applications that are or can be massively parallelised so each little job can be performed separately even if it is slower. The idea is to offload parts of a given job to the GPU and leave the rest to run on the normal CPU. For instance, in a GA, we assign each chromosome to one GPU, and in addition, we can even run several populations simultaneously while obviously collecting their results and analysing them using the CPU core or otherwise. For instance, Pospichal et al. (2010) adopted such a strategy using a high-level language for GPU such as Compute Unified Device Architecture (CUDA) based on the NVIDIA Test platform (Nvidia 2007) when using their parallelised GA on the benchmark functions commonly used in the literature. The usual challenges include the performance of a massive parallelism to make it worthwhile, the good use of shared memory among the multi-processors and obviously the scheme on how to eliminate the system bottleneck. It is also noting that within certain tasks, the use of codes that are written in a parallel way as previously shown can, though a bit tricky and challenging, enhance the speed-up of the whole procedure even further.

6.8   Fuzzy Logic

In practice, the problem may contain some form of vagueness, ambiguity and imprecision with the measurement of the parameters, the conflicting objectives or the constraints. Fuzzy set theory attempts to provide some smoothness around those values that may not be crisp (obviously some are fixed and known with precision etc.). Zadeh (1965) developed the fuzzy set theory where optimisation, for example, works under fuzzy environment. As an example, a person could be seen as not too short but not tall too either. In quality management, for instance, the decision maker highlights that the quality of a given product needs to be categorised as low or good using some crisp threshold. However, to be fair with the decision maker, he/she is likely to consider those products around the threshold if possible. Consider the case where a product is considered poor if quality < A, good if quality > B but questionable if the quality lies between A and B.

A crisp measure would be as follows: poor if quality $< \widehat{Q}$ (say $\widehat{Q} = \frac{(A+B)}{2}$) and good otherwise. In fuzzy logic, we extend the decision around the threshold by using some membership function $F(q) \in [0, 1]$ which can be defined as follows:

$$F(q) = \begin{cases} 0 & \text{if } q < A \text{ (poor quality)} \\ g(A, B, q) & \text{if } A < q < B \text{ (can be either)} \\ 1 & \text{if } q > B \text{ (good quality)} \end{cases}$$

An example of quality where $g(A, B, q)$ can be convex, concave or linear is shown in Fig. 6.3. One possible linear function is given by $g(A, B, q) = \frac{q-A}{B-A}$ where the possible lower (A) and upper (B) levels are set by the decision maker who is happy to make his/her decision crisp outside this range. For example, if the quality falls below the minimum level A, the product is considered poor with 100 % confidence, and similarly if the quality gets over B, it is also seen

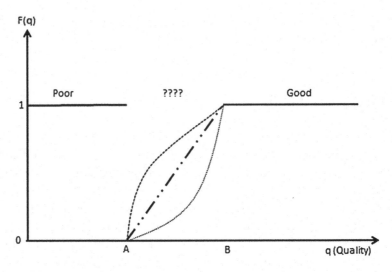

Fig. 6.3 Fuzzy quality

to be good with 100 % confidence. In other words, it is only when it is between these bounds where the problem arises.

Fuzzy logic is based on defining the membership function, the two operators (or and and), and the set of rules such as if then. For example, if we could say if the price of a product is not that expensive, the quality is good and the life cycle of the product is long enough, we can consider such a product to be worth having. Similarly in Logistics system, the selection of the suppliers can be performed by using certain attributes such as quality of product, reliability of the suppliers, their responsiveness, their cost and the degree of relationship. All these attributes can be measured using membership functions which provide a fuzzy score and hence a weighted cumulative score which then provide a rank for each of the suppliers. Fuzzy logic is also extended to multiple and conflicting objectives where the weighted sum approach is usually adopted. This approach is widely used in several applications especially in engineering such as water resource management. For more information on this fascinating subject and the references therein, see Kahraman (2008).

6.9   Dealing with Multiple Objectives

When the problem has several objectives some of which are conflicting, one optimal solution may not exist as the search needs to explore all the various objective functions simultaneously.

Mathematically, the problem can be expressed as follows (case of minimisation):

$$\text{Minimise } Z = F(X) = \left[f_1(X), \ldots, f_p(X) \right]$$
$$\text{such that } X = (X_1, \ldots, X_n) \in S_X; Z = (Z_1, \ldots, Z_p) \in S_Z$$

Where X is an $n-$ dimensional decision vector, F represents the p individual objective functions (i.e., cost, environmental effect, health impact, etc.), whereas S_X and S_Z are feasible regions for the decision variables and the objective functions.

In a single objective problem, two solutions can be easily compared as one of them may produce a better objective function value than the other. However, in the multi-objective scenario this is not the case as one solution can be better in terms of one objective but not the other and vice versa.

We say that a solution X^1 dominate solution X^2 if

$f_k(X^1) \leq f_k(X^2) \forall k = 1, \ldots, p$ and there is at least one $i \in [1, \ldots, p]$ with $f_i(X^1) < f_i(X^2)$

The idea is to determine those solutions that are not dominated by any other ones which are also known as efficient points or Pareto set. Note that there would be <u>no concept of optimality</u> but solution efficiency and the non-dominance property instead. In other words, the idea is to find the set of <u>efficient or non-dominated</u> solutions. These are known as the Pareto points making the Pareto frontier originally discovered by the Italian engineer and economist Vilfredo Pareto over a century ago (Pareto 1896).

The question reduces to how to discard the dominated solutions or to generate the non-dominated ones more intelligently and efficiently. One approach would be to tackle this problem by solving it for each individual objective function separately so to find the 'ideal' solution for each case.

This information can then be used to restart the problem of minimising a new objective function which can be the sum of the weighted deviations between the new and the ideal objective function values. This is similar in principle to the weighted goal programming where the goal is predefined, whereas here it is determined a priory. Another possibility is to solve the problem using several weights in the weighted objective function. If this latter is repeated several times using different set of weights an approximation of the efficient frontier (efficient solutions) can then be constructed. One of the drawbacks is that the number of combinations can be rather large especially if the number of objective functions is more than two besides that each one is in itself a combinatorial optimisation problem which can be hard to solve optimally.

For simplicity, consider, for example, the case of two objective functions $f_1(X)$ and $f_2(X)$ and the heuristic solutions found at a given iteration. For guidance, the Pareto curve which is theoretically obtained by solving the following problem optimally is also shown.

$$\text{Minimising } F(X, \alpha) = \alpha \widetilde{f}_1(X) + (1 - \alpha)\widetilde{f}_2(X)$$

where

$\widetilde{f}_1(.)$ and $\widetilde{f}_2(.)$ denote the normalised values of f_1 and f_2 respectively,

$\alpha \in [0, 1]$ is a parameter reflecting the importance of each of the two functions. Its values start with 0 and goes to 1 with a step size of $\varepsilon = 0.01$ say.

The normalised functions can be defined as $\widetilde{f}_k(X) = \frac{f_k(X)}{\text{Max}\,(f_k(X); X \in \Sigma)}$; $k = 1, 2$ with Σ representing the set of feasible solutions found so far.

If the problem is solved heuristically, the solutions will obviously be dominated by those extreme points that lie on the Pareto curve and the idea is to aim to get closer to this curve by pushing these heuristic solutions towards the curve which is already not given. Figure 6.4 illustrates the Pareto curve (made up of extreme points), the ideal point which is infeasible and the other heuristic solutions that could be enhanced if possible by guiding them towards the direction of the arrows shown in the figure. These solutions could be found using various α values by a given one point-move solution approaches based on several initial solutions

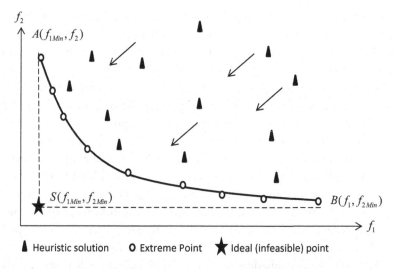

Fig. 6.4 Pareto curve, heuristic solutions and the ideal point

(Chap. 3 or 5), or using one of the evolutionary approaches (Chap. 4 or 5). Most approaches in this area are usually based on the latter and especially focussing on non-linear optimisation where the interest originated from engineering applications.

Note that intelligent and adaptive schemes for generating the appropriate weights to speed up the process from one run to the next do exist, but it is still in its infancy and hence research in this area is challenging but worthwhile.

As some of these combinations (heuristic solutions) are useless or may generate similar solutions when improved by a local search say, one may be tempted to cut on this number but at the expense of missing some of the promising combinations. It is therefore important to provide an appropriate selection mechanism which can be combined with a fast way of sorting the non-dominated solutions. This is especially important in evolutionary techniques which are the one mostly used in this area of multi-objectives. An interesting sorting approach for multi-objective optimisation that incorporates elitism and the construction of successive Pareto fronts is developed by Deb et al. (2002). Here, the first front represents the true approximate Pareto front that is made up of non-dominated solutions only, followed by a second front which contains those dominated solutions by one solution only, then by two and so on. These fronts are constructed based on the following two items. For each solution $X_i; i = 1, \ldots, M$, the number of solutions that dominates it is recorded as n_i and those solutions that are dominated by X_i are stored in E_i. The first front Λ_1 is found by assigning all those solutions that have $n_i = 0$ (i.e., non-dominated solution, current approximate Pareto front), then for each solution in Λ_1 their E_i is adjusted by recording those dominated solutions with one solution only, to make up the second front Λ_2 and so on until all solutions are checked. This DS that uses a systematic book-keeping, though requires slightly more storage from O (M) to $O(M^2)$ with M denoting the number of solutions at that genera-tion, the computation complexity of the sorting has reduced from $O(pM^3)$ to $O(pM^2)$ with p defining the number of objective functions. This fast sorting mechanism embedded with the use of crowding distance (i.e., the largest cuboid) is applied in the selection process at all the stages of the evolutionary technique. In other words, when the selection is between two solutions, the one with a lower rank (the one that belongs to the lower front) is chosen but in case of tie (i.e., both are non-dominated or both with the same n_i) the one with the largest crowding distance is chosen instead.

In terms of selection, one way to guide the local searches is the following: for instance, given a current solution $S_c(f_1, f_2)$, a neighbouring solution $S \in N(S_c)$ is generated either randomly or using a greedy selection rule, see Fig. 6.5. The solution S is accepted depending on whether

(i) S is entirely not dominated (i.e., $f_1(S) < f_1(S_c)$ and $f_2(S) < f_2(S_c)$). In other words, S belongs to the non-dominated quadrant Q_{ND}.

(ii) S is partially not dominated (i.e., $f_1(S) < f_1(S_c)$ or $f_2(S) < f_2(S_c)$ but not necessary both). In other words, S belongs to the two partially non-dominated quadrants such as $Q_{ND1} \cup Q_{ND}$ or $Q_{ND2} \cup Q_{ND}$.

(iii) S is either fully or partially not dominated. In other words, S can be either in Q_{ND}, Q_{ND1} or Q_{ND2}.

(iv) S is dominated (i.e., $f_1(S) \geq f_1(S_c)$ and $f_2(S) \geq f_2(S_c)$). S belongs to the dominated quadrant Q_D.

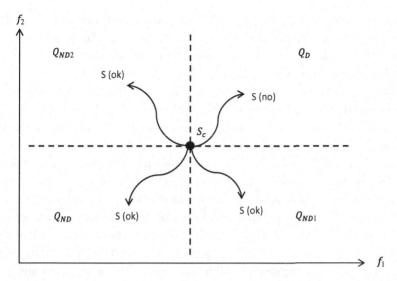

Fig. 6.5 Move selection from S_c to $S \in N(S_c)$ in bi-objective optimisation

The neighbourhood defined by (i) is limited but could lead to efficient points that will contribute to making up the Pareto front, whereas the one described by (iii) is the largest one which could be expensive to explore on its entirety, whereas (ii) acts as a compromise and sits in between the two. The idea is to continue applying the local search on those solutions that are found from either (i) to (iii) with the aim to strengthen the non-dominated Pareto front. The search would terminate when all the neighbouring solutions examined are in the region defined by (iv) (i.e., dominated ones).

Performance Measures The performance measures used to obtain an efficient approximation of the Pareto curve include the number of solutions in the approximated Pareto front, the hyper-volume indicator (*hyp*), besides the usual measures such as the total computational time, the number of evaluations (the number of generations in evolutionary methods) and so on. The first measure that relates to the number of solutions is interesting. At the end of the search, one would like to see a large number of points on the Pareto frontier making the curve as smooth as possible. However, during the search, the spread of these solutions is crucial as niches could be produced that could distort the progress of the search if no special attention is given to these areas. Some attempts of overcoming this drawback are made by several authors who developed niching techniques, see Zitzler and Thiele (1999) and references therein. Among the above performance criteria, it was found that *hyp* is the only measure that is strictly monotonic with respect to Pareto dominance which is extremely important when moving from one approximate Pareto front to a stronger one. This measure was originally presented by Zitzler and Thiele (1999) to define the size of the space covered by the various objective functions from those points that make up the approximate Pareto curve. However, the computation of such hyper-volumes is itself a hard problem to solve especially if the number of objective functions is four or more. The commonly used way was to perform such a task at the very end only to evaluate the approach developed. Very recently, this measure is also incorporated within the search for the selection and the reproduction especially in evolutionary methods as approximation techniques are now developed to reduce the burden of computing such hyper-volumes, see Bader and Zitzler (2010) and references therein.

6.10 Summary

Some key aspects that can enhance the efficiency of a given heuristic implementation are discussed. These include the use of efficient DS, neighbourhood reduction, duplication identification, among others. Due to the computing power and advances in computing, the effect of parallelisation is also covered. As in practice, the input parameters are not crisp a brief on fuzzy logic is given. Also, practical problems may have several objectives some of which are conflicting, ways on how to deal with such an issue are also presented.

References

Adenso-Diaz, B., & Laguna, B. (2006). Fine tuning of the algorithms using fractional experimental designs and local search. *Operations Research, 54*, 99–114.

Bader, J., & Zitzler, E. (2010). HypE: An algorithm for fast hypervolume-based many objective optimization. *Evolutionary Computation, 19*, 45–76.

Deb, K., Agrawal, S., Pratap, A., & Meyarivan, T. (2002). A fast elitist non-dominated sorting genetic algorithms for multi-objective optimization: NSGA-II. *IEEE Transactions on Evolutionary Computation, 6*, 182–197.

Dorigo, M., Caro, G., & Gambardella, L. (1999). Ant algorithms for discrete optimization. *Artificial Life, 5*, 137–172.

Gendreau, M., Hertz, A., & Laporte, G. (1994). A tabu search heuristic for the vehicle routing problem. *Management Science, 40*, 1276–1290.

Grefenstette, J. J. (1986). Optimization of control parameters for genetic algorithms. *IEEE Transactions on Systems, Man, and Cybernetics, 16*, 122–128.

Hopfield, J., & Tank, D. (1985). Neural computation of decisions in optimization problems. *Biological Cybernetics, 52*, 141–152.

Irawan, C. A., Salhi, S., & Drezner, Z. (2016). Hybrid metaheuristics with VNS and exact methods: Application to large unconditional and conditional vertex p-centre problems. *Journal of Heuristics, 22*, 507–537.

Kahraman, C. (2008). *Fuzzy multi-criteria decision making: Theory and applications with recent developments*. London: Springer.

Kelly, J. P., Golden, B., & Assad, A. A. (1993). Large-scale controlled rounding using tabu search with strategic oscillation. *Annals of Operations Research, 41*, 69–84.

NVIDIA, C. (2007). *The open unified device architecture programming guide*. Santa Clara: NVIDIA.

Osman, I. H. (1993). Metastrategy simulated annealing and tabu search algorithms for the vehicle routing problem. *Annals of Operations Research, 41*, 421–451.

Osman, I. H., & Salhi, S. (1996). Local search strategies for the vehicle fleet mix problem. In V. J. Rayward-Smith, I. H. Osman, C. R. Reeves, & G. D. Smith (Eds.), *Modern heuristic search techniques* (pp. 131–154). New York: Wiley.

Pareto, V. (1896). Cours d'Economie Politique: I and II, F. Rouge, Lausanne.

Parhani, B. (2006). *Introduction to parallel processing algorithms and architectures*. London: Springer Science and Business Media.

Pospichal, P., Jaros, J., & Schwarz, J. (2010). Parallel genetic algorithm on the CUDA architecture. In C. Di-Chio et al. (Eds.), *EvoApplications 2010* (Part I, Lecture notes in computer science, Vol. 6024, pp. 442–451). Berlin: Springer.

Rumelhart, D., & McClelland, J. (1986). *Parallel distributed processing*. Cambridge, MA: MIT Press.

Salhi, S. (2002). Defining tabu list size and aspiration criterion within tabu search methods. *Computers and Operations Research, 29*, 67–86.

Salhi, S., & Sari, M. (1997). A Multi-level composite heuristic for the multi-depot vehicle fleet mix problem. *European Journal of Operational Research, 103*, 78–95.

Salhi, S., & Irawan, C. A. (2015). A quadtree-based allocaltion method for a class of large discrete Euclidean location problems. *Computers and Operations Research, 55*, 23–35.

Smithies, R., Salhi, S., & Queen, N. (2004). Adaptive hybrid learning for neural networks. *Neural Computation, 16*, 139–157.

Sze, J. F., Salhi, S., & Wassan, N. (2016). A hybridisation of adaptive variable neighbourhood search and large neighbourhood search: Application to the vehicle routing problem. *Expert Systems with Applications, 65*, 383–397.

Toth, P., & Vigo, D. (2003). The granular tabu search and its applications to the vehicle routing problem. *Informs Journal on Computing, 15*, 333–346.

Woodruff, D. L., & Zemel, E. (1993). Hashing vectors for tabu search. *Annals of Operations Research, 41*, 123–137.

Zadeh, L. A. (1965). Fuzzy sets. *Information and Control, 8*, 338–353.

Zitzler, E., & Thiele, L. (1999). Multiobjective evolutionary algorithms: A comparative case study and the strength Pareto approach. *IEEE Transactions on Evolutionary Computation, 3*, 257–271.

7

Applications, Conclusion and Research Challenges

7.1 Real-Life Applications

Radiotherapy

One of the techniques adopted in treating cancerous tumours is the intensity modulated radiotherapy treatment (IMRT). This consists in sending a dose of radiation to the cancerous region with the aim to sterilise the tumour while avoiding damage to the surrounding healthy organs and tissues. This is performing by defining the number of angles, their respective angles and the intensity chosen for the radiation beams at each of these angles. These are decision variables of a non-linear and highly non-convex optimisation problem. One of the first studies that recognised this challenging optimisation problem was by Stein et al. (1997). Since then the research in this area is extended and new development and automatic tools are commonly used in hospitals. For instance, Bertsimas et al. (2013) present a hybridisation of a gradient descent and an adaptive SA while the initial solution is generated by solving an LP based on using equi-spaced angles beams. Their approach is tested on real-life pancreatic cases (kidneys, liver, stomach, skin and pancreas) at the

© The Author(s) 2017
S. Salhi, *Heuristic Search*, DOI 10.1007/978-3-319-49355-8_7

Massachusetts general hospital of Boston, USA. A case study dealing with patients of head-and-neck tumours at the Portuguese Institute of Oncology in Coimbra is conducted by Dias et al. (2014) who adopted a hybrid GA with neural network. Here, each chromosome is binary and represented by 360 genes, each one for each possible angle, with the angles selected represented by 1. Very recently, Dias et al. (2015) proposed SA, with a dynamically adjusted neighbourhood (in terms of angles) and successfully tested it on the same case study in Coimbra. Their results suggest that a reduced number of angles (and hence less technical adjustment) is required and an improvement in organ sparing and coverage of the tumours is observed.

Sport Management

A variety of metaheuristics have been used to schedule fixtures across many different sporting activities. Wright (1994) produces a computer system that incorporates TS to schedule the English county cricket fixtures. Willis and Terrill (1994) use SA to schedule Australian state cricket. Costa (1995) adopts an evolutionary TS to schedule National Hockey League matches in North America while Thompson (1999) used the same algorithm to schedule the 1999 Rugby World Cup. Wright (2005) produces schedules for New Zealand cricket matches using SA, whereas Kendall (2008) produces a form of local search to schedule English football fixtures over the Christmas period.

Manpower Scheduling

There are many areas in the public and the private sectors where the planning of the people scheduling is complex and time consuming for senior managers. For simplicity, the two areas of educational timetabling and nurse rostering are briefly discussed here.

Educational Timetabling Constructing the timetables including examination timetables at schools, colleges and universities can be a tremendous task if performed manually. Computer systems that incorporate some form of heuristics are nowadays used frequently. For instance, Wright (1996) constructed a tool that incorporates TS for a large comprehensive school of over 1400 pupils and 80 teachers in Lancashire, England. That system helps the school enormously to generate timetables very quickly so they can amend it if necessary instead of spending many hours performing it manually. Dowsland and Thompson (1998) constructed an examination timetable for the University of Swansea using SA. For more information and references therein on educational timetabling, see Burke et al. (2007).

Nurse Rostering This problem plays an important part in efficiently managing the personnel hospital. The aim is to balance the workforce workload while providing flexibility and preferences whenever possible leading to a reduction in stress, an increase in staff satisfaction and a happier working environment. Dowsland and Thompson (2000) integrate ideas from knapsack, network and TS to construct an efficient computer software tool to solve the nurse rostering problem in a large UK hospital in Wales. Burke et al. (2010) adopt the SS approach on real-life instances. For more information on this area, see the review paper by Burke et al. (2004) and the winning paper awarded to Valouxis et al. (2012) by the first international nurse rostering competition (INR2010).

Computational Chemistry

There are many areas within this vast field, but for illustration, I shall briefly mention the studies on Platinum–Palladium clusters (short for P_t $-P_d$). The size of clusters can vary from a small number to millions of atoms and molecules. For example, metals have some practical usefulness in catalysis and nano-electronics. Llyod et al. (2004) adopted a GA to minimise the energy structure which is based on the interaction measured by the Gupta many-body potential. Around 18–20 atoms were used, and it was observed that the P_d atoms have the tendency to be on the surface of the cluster, while the P_t atoms take over the interior site. The structure of a new cluster $P_{t19} - P_{d19}$ was optimised by Llyod et al. (2005) using a powerful GA with special characteristics such as intelligent population seeding, dynamic mutation rates and efficient removal operators. For more information on this subject and references therein, see Ferrando et al. (2008).

Distribution Management (Routing)

Planning the routes by efficiently scheduling the sequence of the customers as they are served and in some cases determining strategically the right vehicle fleet constitutes a massive part in logistic cost (in the range of 30 %), and therefore any improvement gained will provide the company with a competitive edge over its competitors. For instance, Semet and Taillard (1993) used TS to solve a real-life distribution problem in Switzerland leading to about 15 % improvement. Rochat and Semet (1994) developed a TS approach for a pet food company having 100 farms and stores leading to about 16 % cost saving. Brandao and Mercer (1997) also used TS for the multi-trip problem at a British biscuit company in the UK. TA was adopted by Tarantilis and Kiranoudis (2002) to schedule the fresh meat distribution with heterogeneous fleet in a densely populated area of Athens. Tarantilis and Kiranoudis (2007) used a two-phase approach based on LNS for both a dairy and a construction company in Greece. The delivery of blood products to Austrian hospitals for the blood bank of the Austrian Red Cross for Eastern Austria was conducted by Hemmelmayr et al. (2009) using a combination of integer programming and VNS.

Location Issues

There is always a challenge in deciding where to locate something which may require a massive investment such as plant and warehouses, consolidation points and in some cases less expensive equipment that are required in large numbers. For instance, when it comes to locating emergency facilities such as police station, fire station and so on, the aim is to locate the facilities in such a way that the longest time to reach the customer is minimised. This type of problem is known as the p-centre problem where the parameter p can be changed for scenario analysis purposes. Pacheto and Casado (2004) adopted SS to locate a number of geriatric and diabetic health care clinics in the rural area of Burgos in Spain. HS was used by Kaveh and Nasr (2011) to locate bicycle stations in the city of Isfahan, Iran. Lu (2013) implemented SA to locate urgent relief centres in Taiwan to respond to the 7.3 Richter-scale earthquake. There are obviously a large number of applications related to location. As an example, in logistics, one of the dominant drivers is the transportation activity and the less-than-truckload (LTL), in particular. Choosing the right number and the right location for consolidation of small shipment is a strategic issue that could provide a company with the added competitive advantage over its competitors due to its cost efficiency. This question leads to the design of a hub and spoke network. Cunha and Silva (2007) presented an efficient configuration of such a network for one of the top ten LTL trucking companies in Brazil using GA. Also, electricity providers seek to locate their large number of protection devices (costing approximately £10,000 each) on their tree network so to protect the users from having electricity cut in case of big storms and so on. James and Salhi (2000, 2002) explored this unusual network location problem for the Midland Electricity Board (MEB, UK) provider using a constructive heuristic and TS.

Financial Portfolio Management

Whether your portfolio is asset-based or project-based or any other attributes, managing the right mix given each one has its expected profit as well as the risk that it may incur is important. Here, I will highlight the one used in finance. Markowitz (1952) in his finance novel prize paper put forward a mathematical model based on mean-variance. Several complex models that incorporate more than two conflicting objectives have now been studied by metaheuristics mainly using evolutionary algorithms such as GA. For instance, Ehrgott et al. (2004) extended the original model to cater for five objectives (dividend, short and long return, risk, and liquidity) using a hierarchical multi-criteria decision making approach. Woodside-Oriakhi et al. (2011) revisited the original problem by studying the cardinality constraints efficient frontier using GA.

Civil Engineering

The design of water distribution network is very important and costly in the area of civil engineering. This can be seen as a hydraulic infrastructure composed of several pipes of different diameters, hydraulic devices with various powers and different reservoirs. The optimal design of such a problem is to determine the minimal diameter for each pipe in such a way that the total cost is minimised and appropriate water pressure is reached at each of the nodes of the network. HS was tested on this complex non-linear problem based on the water distribution network of Hanoi in Vietnam by Geem et al. (2001). A real coding GA that incorporates neighbourhood reduction with several crossover and mutation operators was proposed by Kadu et al. (2008) for the same case study.

Reservoir management is also one of the key aspects in water resource planning. Each reservoir has several conflicting objectives as well as different operation rules and operating policies due to the land or the cities around it. The aim is to determine the right policy among a large set of possible ones at a given period. For instance, two basic conflicting objectives is the minimisation of the lack of irrigation against the maximisation of the generation of electricity (e.g., hydropower generation). The problem is transformed into a weighted multi-objective approach and solved efficiently using an adaptation of PSO by Kumar and Reddy (2007). The same authors a year earlier put forward an approach based on ACO to solve such multi-purpose reservoir problem, see Kumar and Reddy (2006).

Chemical Engineering

There are many facets in this field but for demonstration purposes, I shall briefly discuss the control of pH in reactors. Such a control is critical in many industries such as pharmaceutical, electro-hydrolysis, wastewater treatment, biotechnology and chemical processes. Regulating the pH value in an acid base titration process that keeps continuously changing is complicated because of neutralisation. pH is considered as an important issue in modelling and has attracted a good amount of research into advanced non-linear control techniques though in practice, the industry still uses linear control techniques mainly due to their simplicity of control and their robustness. Determining suitable parameters values for the control is a challenging task that metaheuristics are called upon. For instance, Mwembeshi et al. (2004a) adopted a GA to intelligently model and control pH in reactors using a lab-scale pH reactor in Birmingham, UK. A similar application that aims to provide a flexible online modelling and control of pH is carried out using neural network by Mwembeshi et al. (2004b). In both studies, a strong base, namely, sodium hydroxide is used to neutralise the process made up of four acids; hydrochloric acid, acetic acid, sulphuric acid and phosphoric acid. Tan et al. (2005) also used a GA to model and control a pilot pH plant focussing on the titration reaction between a week acid such as acetic acid and a strong base like sodium hydroxide. In a similar but slightly different application, Salhi and Fraga (2011) adopted PPA to successfully solve a multi-objective optimisation problem arising in Chemical Engineering. Here, the optimisation relates to a purification section of a process for the production of chlorobenzene. In the overall process, large quantities of benzene are used. Due to the partial conversion in the reaction section of the process, significant quantities of unreacted benzene could be wasted if not recycled. To recycle the benzene, a stream consisting of primarily benzene needs to be further purified to ensure that the benzene sent back upstream in the process is pure enough to not affect the reaction.

Other Engineering Applications

Engineering design problems are usually non-linear and highly constrained. These include among others the pressure vessel design where a hemispherical heads are put at both ends of the cylindrical vessel and the aim is to minimise the sum of the costs due to material, welding and forming. The decision variables can include the thickness of the shell, thickness of the head, its inner radius and the length of the section of the vessel. Due to engineering restriction, the number of available level of thickness is limited to certain values only. This problem has been solved by GA and also by HS. Other problems can be found in Mahdavi et al. (2007). The oil industry has also many applications which include the design of an offshore pipeline network (Brimberg et al. 2003), the pooling problem (Audet et al. 2004) and the scheduling of walkover rigs for Petrobas (Aloise et al. 2006) where VNS is adopted. An interesting review on the use of DE on several engineering problems including multi-sensory fusion, aerodynamic systems, power electronic, filter design among others, is given by Plagianakos et al. (2008). PPA is also used in other industrial applications as described by Sulaiman et al. (2014). There are many applications in other areas of engineering such as electrical engineering, mechanical, environmental and civil engineering where evolutionary methods including GA are commonly used. For example, the edited book by Dasgupta and Michalewicz (2013) can be a useful and informative addition to the reader specifically interested in the use of heuristics with a focus on evolutionary approaches applied to engineering areas.

General/Academic Type Applications

These are general optimisation problems where academics are extensively involved through the design, experiments and analysis in their new heuristics. These applications are so challenging that they are usually used as a platform for PhD theses and basic research. Some of these are also adapted and applied in real-life settings. Hansen et al. (2010) provide an interesting review paper that covers a wide range of these academic-based applications. For completeness, some areas are briefly mentioned here though this is by no means an exhaustive list.

Graphs and networks (maximum clique, constrained minimum spanning tree, max cut, k-cardinality tree, prize-collection Seiner tree colouring, time tabling, etc.),

Vehicle routing (fleet mix, multi-depot, pick and delivery, cumulative, green routing, load balancing, electrical vehicles, packing and routing, location-routing, inventory routing, etc.),

Scheduling (flexible machine scheduling, construction resource scheduling, offshore wind farm installation scheduling, integration of scheduling and transportation, etc.),

Location (reliability in p-median and p-centre, robust location, effect of congestion, constrained continuous, competitive, hub and spoke, integration of location with other supply chain drivers, etc.),

Others (some include data mining, machine and generator maintenance, bioinformatics, health and medicine, environmental issues, financial optimisation, engineering problems, sport management, among others).

7.2 Conclusion

In this monograph, several heuristic-based techniques that are used in solving difficult combinatorial and global optimisation problems are described and their pros and cons highlighted. These range from those that only accept improving solutions as described in Chap. 2 (e.g., hill climbing, VNS, GRASP, multi-level, perturbation, etc.), those that accept non-improving ones while incorporating some form of guidance to avoid the risk of diverging and cycling as presented in Chap. 3 (SA, TA, TS). Those techniques that use simultaneously more than one solution at a time, also known as population or evolutionary methods, are given in Chap. 4 (GA, ACO, Bees, PSO, DE, HS, SS, etc.). As strengths and weaknesses can be found in any heuristic, hybridisation is one way forward to overcome these limitations. This is covered in Chap. 5 with an emphasis on exact with heuristics, heuristics with exact and heuristic with heuristic. A possible hybridisation for data mining that adopts heuristics with increasing complexity as the search progresses is also proposed in this chapter. The efficiency of heuristics depends on several aspects including the hidden key elements when it comes to their implementation. This delicate issue which has a massive practical importance is put across in Chap. 6 where the benefits of constructing effective data structure, neighbourhood reductions, duplication identification schemes among others are highlighted. As real-life problems are not necessarily defined by an input that uses crisp parameter values only, a brief on fuzzy logic is given here. Decision makers may also be faced with conflicting objective functions in practice, and hence this chapter also deals with these more difficult decisions.

In summary, the following guidelines could be helpful when designing a new heuristic:

- Understand the problem and its characteristics.
- Verify whether such a problem or a related one already exists in the literature. If the answer is yes, then review the approaches used and try to modify them accordingly to solve yours. However, if the answer is no, a full and fresh thinking is necessary. Your experience could help you in enhancing and adapting one of the known heuristics or a hybrid which you need to carefully identify for further exploration.
- Try to test your heuristic first on a small example that has the characteristics of the problem. This helps you to see whether the answer makes sense, and to discover any hidden situations that could arise in future and hence validate and even enhance your approach.
- Write an initial implementation even if it is slow and crude. That will serve you as a basis for any enhancement you will subsequently produce. A new solution with better quality or a reduction in computation time will give you a moral boost and as a by-product will encourage you to search for further improvement to your heuristic. This step is usually bypassed as it sounds inefficient and unnecessary, but I believe this is crucial for guaranteeing an efficient heuristic design especially if you happen to be less experienced or a new comer to this area.
- Following the previous step, you can now incorporate what you have learnt to construct any necessary data structure and possible neighbourhood reduction that is worth exploring. It is always a great relief and pleasure when your program runs several times faster without losing the solution quality obtained earlier.
- If the problem you are addressing already exists in the literature, the obvious way to see how good your heuristic works is to compare its performance against the best known methods in terms of both quality and computational time while taking into account the machines used (though coding itself could be an important factor too!). At this time,

avoid any shortcuts for selecting the parameter values that you believe will produce better results. This view is usually short-sighted and does not provide you with the platform for being more inquisitive and critical. However, if no results exist, this becomes tricky. If you are able to formulate the problem mathematically and use a commercial optimisation solver, the obtained LBs and UBs found within a certain time limit, if these are generated, can be useful. Also, if the optimal solutions for small-size instances with similar characteristics to yours exist, that will be a good starting point. The expectation here is that your heuristic will behave similarly when tested on other instances. It may be that your solution is inferior, but remain positive, and use this information to assist you in improving your method (so either way you are a winner!).

7.3   Research Challenges

In this last section of the monograph, I summarise some of the research aspects that excite me and hopefully will do the same for you. I also would like to stress that some of these points have already been briefly highlighted in their respective chapters alongside others which are not repeated here for simplicity.

The first issue, in case you are interested and comfortable with optimisation techniques, is the hybridisation of heuristic and exact methods. This is a nested approach (heuristics within exact methods, or exact methods within heuristics), which, in my opinion, will remain a hot topic for many years to come. However, we may not see the same level of interest in the number of researchers in this particular area as in other areas of heuristic search. This is mainly due to both the level of mathematics and computational skills that are required. The issue here is to dynamically identify those nodes of the B&B tree which are worth examining further. Some of these could have (a) infeasible solutions (i.e., LB values) or (b) feasible ones (UB) that not only include the incumbent solution. Embedding some form of a repair mechanism followed by an improvement scheme, such as local search or even a small version of a powerful heuristic, could be used, for (a) and the improvement scheme for (b). In (a), the aim is to first transform the infeasible solution into a feasible one, whereas in (b), the idea is to improve the feasible solution if possible. Note that the improvement scheme is used not only on the best UB as applied in integer programming for fathoming purposes but also on the other UBs for which their feasible solutions display some form of diversity or promising attributes. This mechanism could lead to improving upon the best incumbent UB.

The view of identifying promising attributes as the search progresses is an exciting avenue that is likely to yield excellent results in the future. The idea is to obtain, through a simpler heuristic which could be called several times such as a smaller version of GRASP, guided LP relaxations, or even evolutionary methods, to generate several solution configurations. These solution configurations when merged together in either subsequent reduced problems or an augmented problem will control the search in a guided manner towards optimal or near optimal solutions due to their promising attributes. This aspect is worth pursuing as shown in various powerful heuristics such as concentration set approach, harmony search, relaxation methods, plant propagation, bucket search, among others.

It can be noted that most heuristics are originally designed with the aim to solve either discrete problems (i.e., TS, VNS, etc.) or non-linear ones (i.e., PSO, DE, HS, etc.). Though there are attempts in adapting these approaches to fit the other type of problem, such as VNS from discrete to continuous, and PSO from continuous to discrete, there is scope for more investigation in this area of transition. For instance, in TS, making a tabu move with respect to discrete points could be represented in the continuous space by a forbidden region defined say by a circle with centre at the new continuous point and a specified radius which decreases as the search progresses. In PSO, on the other hand, moving from a continuous point to a discrete point could be achieved by a simple approximation if possible or through a well-defined mapping that may or may not retain feasibility and so on. Repair mechanisms may need to be incorporated if necessary to regain feasible solutions if the above operators do not include such a feature for efficiency purposes. These are just simple attempts that could be revisited and extended further depending on the characteristics of the problem. It is however worth stressing that such a transformation between the two classes may not be suitable for certain types of problems, or even for some heuristics but this, by no means, is meant to distract those who like the challenge.

In multi-objective optimisation, the search is heavily dependent on a dominance property (in other words, all decisions and searches are linked to non-dominated solutions). This view can be made slightly more flexible by relaxing the approximate Pareto frontier and accepting those dominated solutions that are within a certain threshold. In other words, these solutions that are slightly dominated are not systematically deleted, thus potentially adding extra flexibility as successfully adopted in TA. Another way would be to let the search continue by considering the least dominated solution. This is closely related to the acceptance rule of the least non-improving solution in TS. In both cases, guidance needs to be embedded into the search so to avoid either diverging or cycling. These interesting similarities follow the principles adopted by those techniques that are not restricted to improving moves only (Chap. 3). The reasoning behind the idea is that when applying the local search on those solutions, the new local minima may then become non-dominated ones. For instance, for the threshold-based relaxation, a solution S, though dominated, can be accepted for further investigation if $f_k(S) \leq f_k(S_c) + Th_k$ for a few values of k (say $k \leq p/2$ with p denoting the number of objective functions), with Th_k defining the threshold associated with objective f_k including some with zero threshold. For the TS-based relaxation, some form of tabu definition with respect to dominance needs to be defined.

In this area of multi-objective optimisation, niches may be produced during the search. Their solutions need to be exploited carefully as their contribution, due to sharing, may look less attractive, which, in turn, can be misleading. One way would be to represent each niche by at most m solutions (say $m = 2, 3$ or a fraction of p), each representing one objective function and an additional representative solution defined as the solution of the 1-median problem of the niche.

Also, as highlighted by Bader and Zitzler (2010), the incorporation of hyper-volume as a performance measure can now be used within the search when it comes to selection and reproduction, and not just to evaluate the performance of the technique at the end only. Due to computer power and advances in the development of approximation techniques for the computations of the hyper-volume, it is now relatively more practical to pursue this avenue.

One exciting view in multi-objective optimisation, which I think is promising, is the continuous production of an approximation of a smooth curve based on the current approximate Pareto points. Direct numerical methods can be considered for this purpose. The implementation of this mechanism will depend on the times specified to call for the curve fitting operator, the number of the new approximate points to be considered, the choice of these points, the fitting technique chosen, among others. This construction can be either systematically or periodically performed with the aim of controlling the direction of the search so as to strengthen the approximation of the Pareto curve and get closer to the true one if possible.

Hybridisation of heuristics with heuristics such as MAs is also an interesting avenue of research that has a lot of potential to explore many uncovered grounds. Though a few attempts have been made, most have the tendency to opt for some form of post optimisation either at the end of the search, or systematically or periodically throughout the search. The first is the oldest and the quickest one which is originally used to improve the final solution through the use of a local search or even another metaheuristic. The second and third ones are geared for evolutionary methods where several solutions are present in a population at each generation. The second one is an adaptation of the first by applying the local search/metaheuristic at each generation and for each individual solution. This is time consuming and sometimes not efficient as several solutions are similar. The third one improves on the second by applying the operator at certain generation and solution points. Here, it is good to identify attributes of a solution that would suggest exploring it further through the improvement operator. This can be based on solution quality, its diversity with respect to others, the number of times it is practically acceptable to call such an operator and above all how intensive this operator is used. It is also worth questioning whether the intensification needs to be restrictive to the same stopping criterion for each solution or it can be solution-based as well. Another critical point is whether the improvement operator is based on one local search or a metaheuristic, or would it be more efficient if it could act as an adaptive search operator that behaves like a hyper-heuristic based on metaheuristics as highlighted in Chap. 5. These are just some of the points that are likely to influence the overall search strategy of MAs in general, and any other hidden key that shows promise is worth examining.

It is vital to understand how a given method works, and then to design a data structure that incorporates as much as possible those useful intermediate calculations that can avoid recomputing intermediate data. Though this process may increase some level of memory, and may require an initial fixed cost in terms of development and computation time, it does often lead to a massive time saving without affecting the solution quality at all. This time saving aspect can also be enhanced further by the construction of effective neighbourhood reduction schemes that help to avoid even checking combinations and operations that are unlikely to result in improving the solution. A note of caution here is that although the latter schemes may result in a considerable saving in computational time and are usually simple to construct, they could affect the solution quality if they happen to be too restrictive. The compromise in the design between a neighbourhood reduction which is powerful enough (remove as many as possible irrelevant checks) while not excluding promising moves is exciting, has practical implications and hence is worth exploring.

Evolutionary heuristics, as those described in Chap. 4, are relatively easy to parallelise and hence, can be used for larger instances and in practical settings if the computing facilities are available. The integration of GPU and CPU, as briefly highlighted in Chap. 6, can be a useful addition to the mix and may entice researchers, mainly in the computer science community, to be more involved. Though this is a technical issue, it may require some interesting developments that could massively influence the way we implement these techniques.

The design of adaptive search that dynamically learns and makes use of the obtained information is crucial. This learning mechanism which ought to be continuously or at least periodically updated is then used to pseudo-randomly select at each iteration (or at a well-defined iteration, or just after a certain number of iterations) the decision rules to be used. These can include a subset of neighbourhoods to choose from, a number of local searches to be used for intensification purposes or even the powerful heuristics or exact methods to opt from. This kind of search is self-adaptive and also efficient as it uses only what it needs rather than all avenues where there is the expectation that a good solution may be found! This challenging and practical research issue could probably become one of the most popular research areas in the near future as it has the additional benefits of being applicable in several areas ranging from engineering to medicine.

In a related but different aspect, it is well known that most powerful heuristics seem to suffer from parameter tuning. In certain techniques, some parameters may be defined analytically, thus removing their risk effect drastically. In other words, if a parameter is found to have a clear meaning and can be explicitly expressed, that will make the technique more deterministic and easier to implement though it may not always outperform the use of a specific value found after several trials. The above approach may not be easy but is certainly worth the try. Another approach is to concentrate on schemes that incorporate ways of reducing the number of parameters or adjusting the parameters values dynamically and adaptively. This is very welcome and, in my view, is also one of the ways forward. Results from such studies will provide us with tools on how to avoid the excessive time required for fine tuning of the parameters, besides making the heuristic less sensitive to parameter values, and hence more reliable to use. This reliability issue can lead to serious practical limitations and a massive negative publicity which we all would like to live without!

It has to be noted that advances in heuristic search which could lead to solving more complex and realistic problems will only be applicable if an interest in practical problems and the challenges that they bring is achieved, the advances in computer technology and commercial optimisation software are constantly brought to our attention, and above all the theoretical developments in the area of combinatorial and global optimisation are well understood.

Heuristics have evolved from a variety of applications, from different people's expertise and sometimes just as a by-product of curiosity of some researchers whose original aim was to disapprove their usefulness, and who fortunately found themselves trapped and even hooked! I believe this less-structured area, known by some as a grey research area, will remain for many years to come <u>even greyer and open to more challenges</u> that not only enrich the area and its prospects, but also make it a most suitable and attractive optimisation framework for tackling complex combinatorial and global optimisation problems.

References

Aloise, D. J., Aloise, D., Rocha, C. T. M., Ribeiro, J. C., & Moura, L. S. S. (2006). Scheduling workover rigs for onshore oil production. *Discrete Applied Mathematics, 154,* 695–702.

Audet, C., Brimberg, J., Hansen, P., & Mladenović, N. (2004). Pooling problem: Alternate formulation and solution methods. *Management Science, 50,* 761–776.

Bader, J., & Zitzler, E. (2010). HypE: An algorithm for fast hypervolume-based many objective optimization. *Evolutionary Computation, 19,* 45–76.

Bertsimas, D., Cacchiani, V., Craft, D., & Nohadani, O. (2013). A hybrid approach to beam angle optimization in intensity-modulated radiation therapy. *Computers and Operations Research, 40,* 2187–2197.

Brandao, J., & Mercer, A. (1997). A tabu search heuristic for the multiple-trip vehicle routing and scheduling problem. *European Journal of Operational Research, 100,* 180–191.

Brimberg, J., Hansen, P., Lih, K. W., Mladenović, N., & Breton, M. (2003). An oil pipeline design problem. *Operations Research, 51,* 228–239.

Burke, E. K., De Causmaecker, P., Vanden Berghe, G., & Van Landeghem, H. (2004). The state of the art of nurse rostering. *Journal of Scheduling, 7,* 441–499.

Burke, E. K., McCollum, B., Meisels, A., Petrovic, S., & Qu, R. (2007). A graph-based hyper-heuristic for educational timetabling problems. *European Journal of Operational Research, 176,* 177–192.

Burke, E. K., Curtois, T., Qu, R., & Vanden Berghe, G. (2010). A scatter search methodology for the nurse rostering problem. *The Journal of the Operational Research Society, 61,* 1667–1679.

Costa, D. (1995). An evolutionary tabu search algorithm and the NHL scheduling problem. *INFOR: Information Systems and Operational Research, 33,* 161–178.

Cunha, C. B., & Silva, M. E. (2007). A genetic algorithm for the problem of configuring a hub-and-spoke network for a LTL trucking company in Brazil. *European Journal of Operational Research, 179,* 747–758.

Dasgupta, D., & Michalewicz, Z. (Eds.). (2013). *Evolutionary algorithms in engineering applications.* New York: Springer.

Dias, J., Rocha, H., Ferreira, B., & de Carmo, L. C. (2014). A genetic algorithm with neural network fitness function evaluation for IMRT beam angle optimization. *Central European Journal of Operations Research, 22,* 431–455.

Dias, J., Rocha, H., Ferreira, B., & de Carmo, L. C. (2015). Simulated annealing applied to IMRT beam angle optimization: A computational study. *Physica Medica, 31,* 747–756.

Dowsland, K. A., & Thompson, J. M. (1998). A robust simulated annealing based examination timetabling system. *Computers and Operations Research, 25,* 637–648.

Dowsland, K. A., & Thompson, J. M. (2000). Solving a nurse scheduling problem with knapsacks, network and tabu search. *The Journal of the Operational Research Society, 51,* 825–833.

Ehrgott, M., Klamroth, K., & Schwehm, C. (2004). An MCDM approach to portfolio optimization. *European Journal of Operational Research, 155,* 752–770.

Ferrando, R., Jellinek, J., & Johnston, R. L. (2008). Nanoalloys: From theory to applications of alloy clusters and nanoparticles. *Chemical Reviews, 108,* 846–910.

Geem, Z. W., Kim, J. H., & Loganathan, G. V. (2001). A new heuristic optimization algorithm: Harmony search. *Simulation, 76*(2), 60–68.

Hansen, P., Mladenović, N., Brimberg, J., & Moreno Perez, J. A. (2010). Variable neighbourhood search. In M. Gendreau & J. Y. Potvin (Eds.), *Handbook of metaheuristics* (pp. 61–86). London: Springer.

Hemmelmayr, V., Doerner, K. F., Hartl, R. F., & Savelsbergh, M. W. P. (2009). Delivery strategies for blood products supplies. *Operations Research Spectrum, 31,* 707–725.

James, C., & Salhi, S. (2000). The location of protection devices on electrical tree networks: A heuristic approach. *The Journal of the Operational Research Society, 51,* 959–970.

James, C., & Salhi, S. (2002). A tabu search heuristic for the location of multi type protection devices on electrical tree networks. *Journal of Combinatorial Optimization, 6*, 81–98.

Kadu, M. S., Gupta, R., & Bhave, P. (2008). Optimal design of water networks using a modified genetic algorithm with reduction in search space. *Journal of Water Resources Planning and Management, 134*, 147–160.

Kaveh, A., & Nasr, H. (2011). Solving the conditional and unconditional p-centre problem with modified harmony search: A real case study. *Scientia Iranica, 4*, 867–877.

Kendall, G. (2008). Scheduling English football fixtures over holiday periods. *The Journal of the Operational Research Society, 59*, 743–755.

Kumar, D. N., & Reddy, M. J. (2006). Ant colony optimization for multi-purpose reservoir operation. *Water Resources Management, 20*, 879–898.

Kumar, D. N., & Reddy, M. J. (2007). Multi-purpose reservoir operation using particle swarm optimization. *Journal of Water Resources Planning and Management, 133*, 192–201.

Llyod, L. D., Johnston, R. L., Salhi, S., & Wilson, N. T. (2004). Theoretical investigation of isomer stability in platinum-palladium nanoalloy clusters. *Journal of Materials Chemistry, 14*, 1691–1704.

Llyod, L. D., Johnston, R. L., & Salhi, S. (2005). Strategies for increasing the efficiency of a genetic algorithm for the structural optimization of nannalloy clusters. *Journal of Computational Chemistry, 26*, 1069–1078.

Lu, C. (2013). Robust weighted vertex p-center model considering uncertain data: An application to emergency management. *European Journal of Operational Research, 230*, 113–121.

Mahdavi, M., Fesanghary, M., & Damangir, E. (2007). An improved harmony search algorithm for solving optimization problems. *Applied Mathematics and Computation, 188*, 1567–1579.

Markowitz, H. M. (1952). Portfolio selection. *Journal of Finance, 7*, 77–91.

Mwembeshi, M. M., Kent, C. A., & Salhi, S. (2004a). A genetic algorithm based approach to intelligent modelling and control of pH reactors. *Computers and Chemical Engineering, 28*, 1743–1757.

Mwembeshi, M. M., Kent, C. A., & Salhi, S. (2004b). Flexible on-line modelling and control of pH in waste neutralisation reactors. *Chemical Engineering and Technology Journal, 27*, 130–138.

Pacheto, J. A., & Casado, S. (2004). Solving two location models with few facilities by using a hybrid heuristic: A real health resources case. *Computers and Operations Research, 32*, 3075–3091.

Plagianakos, V. P., Tasoulis, D. K., & Vrahatis, M. N. (2008). A review of major application areas of differential evolution. In U. K. Chakraborty (Ed.), *Advances in differential evolution* (Studies in computational intelligence, Vol. 143, pp. 197–238). Berlin: Springer.

Rochat, Y., & Semet, F. (1994). A tabu search approach for delivering pet food and flour in Switzerland. *The Journal of the Operational Research Society, 45,* 1233–1246.

Salhi, A., & Fraga, E. S. (2011). Nature-inspired optimisation approaches and the new plant propagation algorithm. In *Proceedings of the ICeMATH2011,* pp. K2–1 to K2–8. Java, Indonesia.

Semet, F., & Taillard, E. D. (1993). Solving real-life vehicle routing problem efficiently using tabu search. *Annals of Operations Research, 41,* 469–488.

Stein, J., Mohan, R., Wang, X. H., Bortfeld, T., Wu, Q., Preiser, K., Ling, C. C., & Schlegel, W. (1997). Number and orientation of beams in intensity-modulated radiation treatments. *Medical Physics, 24,* 149–160.

Sulaiman, M., Salhi, A., & Selamoglu, B. I. (2014). A plant propagation algorithm for constrained engineering optimisation problems. *Mathematical Problems in Engineering,* pp. 1–17, Hindawi Publishing Corp.

Tan, W. W., Lu, F., Loh, A. P., & Tan, K. C. (2005). Modelling and control of a pilot pH plant using genetic algorithm. *Engineering Applications of Artificial Intelligence, 18,* 485–494.

Tarantilis, C. D., & Kiranoudis, C. T. (2002). Distribution of fresh meat. *Journal of Food Engineering, 51,* 85–91.

Tarantilis, C. D., & Kiranoudis, C. T. (2007). A flexible adaptive memory-based algorithm for real-life transportation operations: Two case studies from diary and construction sector. *European Journal of Operational Research, 179,* 806–822.

Thompson, J. (1999). Kicking timetabling problems into touch. *OR Insight, 12,* 7–15.

Valouxis, G., Gogos, C., Goulas, G., & Alefragis, P. (2012). A systematic two phase approach for the nurse rostering problem. *European Journal of Operational Research, 219,* 425–433.

Willis, R., & Terrill, B. (1994). Scheduling the Australian state cricket season using simulated annealing. *The Journal of the Operational Research Society, 45,* 276–280.

Woodside-oriakhi, C., Lucas, C., & Beasley, J. (2011). Heuristic algorithms for the cardinality constrained efficient frontier. *European Journal of Operational Research, 213,* 538–550.

Wright, M. (1994). Timetabling county cricket fixtures using a form of tabu seach. *The Journal of the Operational Research Society, 45,* 758–770.

Wright, M. (1996). School timetabling using heuristic search. *The Journal of the Operational Research Society, 47,* 347–357.

Wright, M. (2005). Scheduling fixtures for New Zealand cricket. *IMA Journal of Management Mathematics, 16,* 99–112.

Index

Printed in the United States
By Bookmasters